Global Perspectives on Gender and Space

Feminism has re-shaped the way we think about equality, power relations and social change. Recent feminist scholarship has provided new theoretical frameworks, methodologies and empirical analyses of how gender and feminism are situated within the development process. *Global Perspectives on Gender and Space: Engaging feminism and development* draws upon this framework to explore the effects of globalization on development in diverse geographical contexts. It explores how women's and men's lives are gendered in specific spaces as well as across multiple landscapes.

Traveling from South Asia to sub-Saharan Africa to North America and the Caribbean, the contributions illustrate the link between gender and global development, including economic livelihoods, policy measures and environmental change. Divided into three sections, *Global Perspectives on Gender and Space* showcases the following issues: (1) the impact of neoliberal policies on transnational migration, public services and microfinance programs; (2) feminist and participatory methodologies employed in the evaluation of land use, women's cooperatives and liberation struggles; and (3) gendered approaches to climate change, natural disasters and conservation in the global South. A feminist lens is the common thread throughout these sections that weaves gender into the very fabric of everyday life, providing a common link between varied spaces around the globe by mapping gendered patterns of power and social change.

This timely volume provides geographic comparisons and case studies to give empirically informed insights on processes and practices relevant to feminism and development. It illustrates ways to empower individuals and communities through transnational struggles and grass-roots organizations, while emphasizing human rights and gender equity, and will be of interest to those studying Geography, Development Studies, International Relations and Gender Studies.

Ann M. Oberhauser is Professor of Geography at West Virginia University, USA.

Ibipo Johnston-Anumonwo is Professor of Geography at State University of New York College at Cortland, USA.

Routledge studies in human geography

This series provides a forum for innovative, vibrant, and critical debate within Human Geography. Titles will reflect the wealth of research which is taking place in this diverse and ever-expanding field. Contributions will be drawn from the main sub-disciplines and from innovative areas of work which have no particular sub-disciplinary allegiances.

Published:

1 **A Geography of Islands**
Small island insularity
Stephen A. Royle

2 **Citizenships, Contingency and the Countryside**
Rights, culture, land and the environment
Gavin Parker

3 **The Differentiated Countryside**
Jonathan Murdoch, Philip Lowe, Neil Ward and Terry Marsden

4 **The Human Geography of East Central Europe**
David Turnock

5 **Imagined Regional Communities**
Integration and sovereignty in the global south
James D. Sidaway

6 **Mapping Modernities**
Geographies of Central and Eastern Europe 1920–2000
Alan Dingsdale

7 **Rural Poverty**
Marginalisation and exclusion in Britain and the United States
Paul Milbourne

8 **Poverty and the Third Way**
Colin C. Williams and Jan Windebank

9 **Ageing and Place**
Edited by Gavin J. Andrews and David R. Phillips

10 **Geographies of Commodity Chains**
Edited by Alex Hughes and Suzanne Reimer

11 **Queering Tourism**
Paradoxical performances at Gay
Pride parades
Lynda T. Johnston

12 **Cross-Continental Food
Chains**
*Edited by Niels Fold and Bill
Pritchard*

13 **Private Cities**
*Edited by Georg Glasze, Chris
Webster and Klaus Frantz*

14 **Global Geographies of
Post-Socialist Transition**
Tassilo Herrschel

15 **Urban Development in
Post-Reform China**
*Fulong Wu, Jiang Xu and
Anthony Gar-On Yeh*

16 **Rural Governance**
International perspectives
*Edited by Lynda Cheshire,
Vaughan Higgins and
Geoffrey Lawrence*

17 **Global Perspectives on Rural
Childhood and Youth**
Young rural lives
*Edited by Ruth Panelli,
Samantha Punch, and
Elsbeth Robson*

18 **World City Syndrome**
Neoliberalism and
inequality in Cape Town
David A. McDonald

19 **Exploring Post-Development**
Aram Ziai

20 **Family Farms**
*Harold Brookfield and
Helen Parsons*

21 **China on the Move**
Migration, the state, and
the household
C. Cindy Fan

22 **Participatory Action Research
Approaches and Methods**
Connecting people,
participation and place
*Edited by Sara Kindon, Rachel
Pain and Mike Kesby*

23 **Time-Space Compression**
Historical geographies
Barney Warf

24 **Sensing Cities**
Monica Degen

25 **International Migration and
Knowledge**
*Allan Williams and Vladimir
Baláž*

26 **The Spatial Turn**
Interdisciplinary perspectives
*Edited by Barney Warf and
Santa Arias*

27 **Whose Urban Renaissance?**
An international comparison of
urban regeneration policies
*Edited by Libby Porter and
Katie Shaw*

28 **Rethinking Maps**
*Edited by Martin Dodge,
Rob Kitchin and Chris Perkins*

29 **Rural–Urban Dynamics**
Livelihoods, mobility and
markets in African and
Asian frontiers
*Edited by Jytte Agergaard,
Niels Fold and
Katherine V. Gough*

30 **Spaces of Vernacular Creativity**
Rethinking the cultural economy
Edited by Tim Edensor, Deborah Leslie, Steve Millington and Norma Rantisi

31 **Critical Reflections on Regional Competitiveness**
Gillian Bristow

32 **Governance and Planning of Mega-City Regions**
An international comparative perspective
Edited by Jiang Xu and Anthony G.O. Yeh

33 **Design Economies and the Changing World Economy**
Innovation, production and competitiveness
John Bryson and Grete Rustin

34 **Globalization of Advertising**
Agencies, cities and spaces of creativity
James R. Faulconbridge, Peter J. Taylor, Jonathan V. Beaverstock and Corinne Nativel

35 **Cities and Low Carbon Transitions**
Edited by Harriet Bulkeley, Vanesa Castán Broto, Mike Hodson and Simon Marvin

36 **Globalization, Modernity and the City**
John Rennie Short

37 **Climate Change and the Crisis of Capitalism**
A chance to reclaim self, society and nature
Edited by Mark Pelling, David Manual Navarette and Michael Redclift

38 **New Economic Spaces in Asian Cities**
From industrial restructuring to the cultural turn
Edited by Peter W. Daniels, Kong Chong Ho and Thomas A. Hutton

39 **Landscape and the Ideology of Nature in Exurbia**
Green sprawl
Edited by Kirsten Valentine Cadieux and Laura Taylor

40 **Cities, Regions and Flows**
Edited by Peter V. Hall and Markus Hesse

41 **The Politics of Urban Cultural Policy**
Global perspectives
Edited by Carl Grodach and Daniel Silver

42 **Ecologies and Politics of Health**
Edited by Brian King and Kelley Crews

43 **Producer Services in China**
Economic and urban development
Edited by Anthony G.O. Yeh and Fiona F. Yang

44 **Locating Right to the City in the Global South**
Tony Roshan Samara, Shenjing He and Guo Chen

45 **Spatial-Economic Metamorphosis of a Nebula City**
Schiphol and the Schiphol region during the 20th century
Abderrahman El Makhloufi

46 **Learning Transnational
Learning**
*Edited by Åge Mariussen and
Seija Virkkala*

47 **Migration, Risk, and
Uncertainty**
*Allan Williams and
Vladimir Baláž*

48 **Global Perspectives on Gender
and Space**
Engaging feminism and
development
*Edited by Ann M. Oberhauser
and Ibipo Johnston-Anumonwo*

49 **Fieldwork in the Global
South**
Ethical challenges and dilemmas
Edited by Jenny Lunn

Forthcoming:

50 **Intergenerational Space**
*Edited by Robert Vanderbeck
and Nancy Worth*

51 **Performativity, Politics, and
the Production of Social Space**
*Edited by Michael R. Glass and
Reuben Rose-Redwood*

Global Perspectives on Gender and Space

Engaging feminism and development

Edited by Ann M. Oberhauser and
Ibipo Johnston-Anumonwo

LONDON AND NEW YORK

First published 2014
by Routledge
2 Park Square, Milton Park, Abingdon, Oxon OX14 4RN

and by Routledge
52 Vanderbilt Avenue, New York, NY 10017

First issued in paperback 2020

Routledge is an imprint of the Taylor & Francis Group, an informa business

British Library Cataloguing in Publication Data
A catalogue record for this book is available from the British Library

Library of Congress Cataloging in Publication Data
Oberhauser, Ann M.
Global perspectives on gender and space : engaging feminism and development/
Ann M. Oberhauser, Ibipo Johnston-Anumonwo.

    pages cm. – (Routledge studies in human geography)
    Includes bibliographical references and index.
    1. Women in development. 2. Feminism. 3. Sex role.
    I. Johnston-Anumonwo, Ibipo. II. Title.
HQ1240.O245 2014
305.42–dc23   2013024928

ISBN 13: 978-0-367-66959-1 (pbk)
ISBN 13: 978-0-415-65798-3 (hbk)

Typeset in Times New Roman
by Sunrise Setting Ltd, Paignton, UK

Contents

List of illustrations	xi
List of contributors	xiii
Preface	xv
Acknowledgments	xvii

**Introduction: engaging feminism and development –
worlds of inequality and change**   1

IBIPO JOHNSTON-ANUMONWO AND ANN M. OBERHAUSER

**PART I
Feminist perspectives on neoliberal globalization**   15

**1 Gender equity and commercialization of public toilet
services in Nairobi, Kenya**   17

JEREMIA N. NJERU, IBIPO JOHNSTON-ANUMONWO AND SAMUEL O. OWUOR

**2 "Out of the kitchen": gender, empowerment and
microfinance programs in Sri Lanka**   35

SEELA ALADUWAKA AND ANN M. OBERHAUSER

**3 Neoliberalization, gender and the rise of the diaspora
option in Jamaica**   53

BEVERLEY MULLINGS

**4 Stuck in a groove? Gender, politics and globalization in
anti-sex trafficking policy initiatives**   71

VIDYAMALI SAMARASINGHE

PART II
Gendering the field: participatory feminist research   85

5 **Crossing boundaries: transnational feminist
   methodologies in the global North and South**   87

   ANN M. OBERHAUSER

6 **Gender and land use in KwaZulu-Natal, South Africa:
   a qualitative methodological approach**   103

   HUMAYRAH BASSA, URMILLA BOB AND SUVESHNEE MUNIEN

7 **Participatory mapping of women's daily lives:
   perspectives from rural Uganda**   122

   DEBORAH NAYBOR AND RAM ALAGAN

8 **Mapping differential geographies: women's
   contributions to the liberation struggle in Tanzania**   138

   MARLA JAKSCH

PART III
Gender, the environment and community-based development   161

9 **Gender, livelihoods and the construction of climate
   change among Masai pastoralists**   163

   ELIZABETH EDNA WANGUI

10 **Gender mapping in post-disaster recovery: lessons from
    Sri Lanka's tsunami**   181

    RAM ALAGAN AND SEELA ALADUWAKA

11 **Ecodevelopment, gender and empowerment:
    perspectives from India's Protected Area communities**   200

    RUCHI BADOLA, MONICA V. OGRA AND SHIVANI C. BARTHWAL

    Index   224

Illustrations

Figures

1.1   Location of Ikotoilets in Nairobi, Kenya.   24
1.2   Aga Khan Walk Ikotoilet.   25
1.3   Kawangware-Congo Ikotoilet.   26
1.4   Inside the male compartment of the Aga Khan Walk Facility.   29
2.1   Study sites in Kandy District, Sri Lanka.   40
2.2   Women receiving loans at the Samurdhi Bank.   45
5.1   Appalachia, United States and Limpopo Province, South Africa.   90
5.2   Members of the Tshandama Community Project.   95
5.3   (a) Knitter's work space; (b) Knitter's home in rural West
      Virginia.   96
5.4   Training session with knitters.   98
6.1   Map of South African study area—Inanda, KwaZulu-Natal.   107
6.2   Women's perceptions regarding current and future land use in
      Inanda.   113
6.3   Men's perceptions regarding current and future land use in
      Inanda.   115
6.4   Actual land use compared to participant perceptions of land use
      in Inanda.   116
6.5   Women's group Venn diagram.   117
6.6   Men's group Venn diagram.   118
7.1   Case study area showing GPS data points.   127
7.2   Participatory map showing spatial relationship concepts.   130
7.3   GPS tracker paths of multiple participants.   132
7.4   Meeting for identification of GPS locations on GIS maps.   133
8.1   Screenshot of the "mapping" of Bibi Titi Mohamed.   140
8.2   *Khanga* with political message *Umoja wa Wanawake wa
      Tanzania* (Union of Women in Tanzania).   146
8.3   Soweto Historical GIS Project team with local informants using
      ArcGIS on iPad.   150
8.4   Road named after Bibi Titi Mohammed in Dar es Salaam,
      Tanzania.   153

8.5	Virtual Freedom Trail Project (VFTP) map.	154
9.1	Case study sites in Kajiado South Constituency, Kenya.	166
9.2	Irrigated farming in Mbirikani Group Ranch.	170
9.3	Water tap located along the Nolturesh pipeline.	175
10.1	Post-tsunami disaster rehabilitation and recovery study areas in Sri Lanka.	187
10.2	Female perception of shelter relocation, livelihoods and service development.	191
10.3	Male perception of shelter relocation, livelihoods and service development.	192
11.1	Selected protected areas and biogeographic zones in India.	204
11.2	Ladakhi woman picking apricots.	209
11.3	Nimaling pasture at Hemis National Park.	210
11.4	Entry gate at Bhundyar Valley, Nanda Devi Biosphere Reserve.	213

Tables

2.1	Profiles of selected microfinance programs	39
2.2	Educational level of respondents	42
2.3	Employment among borrowers and non-borrowers	43
2.4	Mobility and decision-making among borrowers and non-borrowers	44
6.1	Ranking exercise on land issues in Inanda	111
7.1	Summary of modified Human Development Index variables (as a percentage of participants)	131
9.1	Gender and livelihood uses of plants negatively impacted by climate change in Kajiado South, Kenya	172
10.1	Women's and men's rehabilitation and recovery plans in Kinniya, Sri Lanka	194
11.1	Socio-economic and geographic characteristics of selected Protected Areas in India	206
11.2	Overview of ecodevelopment initiatives in selected Protected Areas in India	207
11.3	Impacts of ecodevelopment in selected Protected Areas in India	208

Contributors

Seela Aladuwaka is Assistant Professor of Geography at Alabama State University. She has conducted research on poverty and microcredit, social impact assessment and gender and natural disasters, with a specific focus in South Asia. She was also a recipient of a Fulbright scholarship.

Ram Alagan is Assistant Professor of Geography at Alabama State University. His research and teaching interests include PGIS, EIA, Resource Management and Disaster Management Studies. He has done research in Sri Lanka and the US and was a Fulbright scholarship recipient.

Ruchi Badola works as a Senior Professor at the Wildlife Institute of India, with the Department of Eco development Planning and Participatory Management. She specializes in community/stakeholder participation in biodiversity conservation, ecological economics assessment of ecosystem services, sustainable livelihood and gender issues in conservation.

Shivani C. Barthwal is Senior Researcher at the Wildlife Institute of India. She studies human-wildlife interaction, resource extraction and distribution, ecosystem services and policy actions in conservation. She participated in the Millennium Ecosystem Assessment as the author for the MA synthesis report.

Humayrah Bassa is a PhD candidate at the University of KwaZulu-Natal, South Africa, in the Discipline of Geography, School of Agriculture, Earth and Environmental Sciences. She conducts research on environmental resource economics and land cover change using remote sensing.

Urmilla Bob is Professor of Geography at the University of KwaZulu-Natal, South Africa, in the School of Agriculture, Earth and Environmental Sciences. She conducts research on a range of development and environmental issues in African contexts, with a specific focus on sustainable livelihoods and gender dimensions.

Marla Jaksch is Assistant Professor of Women's and Gender Studies at The College of New Jersey. She has conducted research and has taught and worked in development in East Africa for almost a decade, most recently as a Fulbright scholar in the Institute of Development Studies at the University of Dar es Salaam, Tanzania.

Ibipo Johnston-Anumonwo, Professor of Geography at SUNY Cortland, has published on gender and employment access. She received a Chancellor's Award for Excellence in Scholarship, and serves on the editorial board of *WAGADU: A Journal of Transnational Women's and Gender Studies.*

Beverley Mullings, Associate Professor of Geography and Gender Studies at Queen's University, is past editor of *Gender, Place and Culture*. Her research focuses on globalization, gender transformations in work and social justice in the global south.

Suveshnee Munien is a PhD candidate at the University of KwaZulu-Natal, South Africa, in Geography in the School of Agriculture, Earth and Environmental Sciences. She is a GIS and remote sensing specialist, whose current research focuses on sustainable livelihoods in marginalized contexts.

Deborah A. Naybor is Adjunct Professor of Geography at the University of Buffalo and executive director of Both Your Hands, a non-profit organization working on economic development in communities around the world. She has conducted research on sub-Saharan Africa and Asia on gendered land rights and women's economic development.

Jeremia N. Njeru is Assistant Professor of Geography at West Virginia University. He has conducted research on urban development, urban environmental problems and social movements with a focus on Kenya.

Ann M. Oberhauser is Professor of Geography at West Virginia University. She has conducted research on gender, rural livelihoods and globalization in sub-Saharan Africa and Appalachia, with funding from the National Science Foundation and the US Department of Agriculture.

Monica V. Ogra is Associate Professor of Environmental Studies at Gettysburg College. Dr Ogra holds a PhD in Geography, with a specialization in gender, environment and development, and maintains a particular interest in wildlife conservation and protected areas. She is also a member of her college's programs in Globalization Studies and Women, Gender and Sexuality Studies.

Samuel O. Owuor is Senior Lecturer at the Department of Geography and Environmental Studies, University of Nairobi, Kenya. His research interests and experience revolves around broad and specific issues in urban development, governance, urban poverty and livelihoods and food security.

Vidyamali Samarasinghe is Professor of Geography at American University. She has conducted research on poverty, rural farming in Sri Lanka and sex trafficking. Her book, *Female Sex Trafficking in Asia*, examines human trafficking from the perspective of rising demand from the global North.

Elizabeth Edna Wangui is Associate Professor in the Department of Geography at Ohio University. Her research focuses on pastoralist communities within the contexts of feminist political ecology, environmental change and development practice, and has been funded by the National Science Foundation.

Preface

In the spirit of transnational feminism, this edited collection transcends boundaries and involves collaborative efforts among colleagues from diverse yet overlapping experiences and intellectual approaches. Contributors to this book have engaged in discussions and developed research agendas across spatial boundaries and in some cases, over decades of academic collaboration. This background not only enriches our perspectives but highlights the possibilities of a feminist project in a global context of educational institutions and applied research.

This project stems from long-term collaboration between the co-editors, who initially met and formed a friendship as students at Clark University's Graduate School of Geography during the 1980s. The "transnational" relationship that developed from this friendship brought opportunities to share personal and academic experiences in their respective countries of the United States and Nigeria. These connections have also been instrumental in forging links that enhance their feminist lens and give a better understanding of the material and ideological aspects of their global perspectives. The development of this book project grew out of various activities and discussions that involved the authors on several occasions. In 2011, many of the contributors to this book participated in a workshop at West Virginia University in order to formulate themes and discuss their research strategies. The following year, they presented their work on gender, development and globalization at two sessions of the Association of American Geographers (AAG) conference in New York City.

This collection of chapters is grounded in, and contributes to, fields of study that include feminist geography, development studies and global studies, and is designed to recognize those who struggle to make the world a more just and liberating place. Contributing authors represent a range of early-, mid- and advanced-career scholars and activists who have made significant impacts in their subject areas and research activities. Specifically, they share academic interests in the field of gender and development, have experience in teaching and conducting research in both domestic and international contexts and are passionate about the possibilities of transnational feminism.

Furthermore, these authors are from diverse world regions that include South Asia, the Caribbean, North America and Africa—where they have conducted primary research on issues concerning neoliberal economics, resource management, climate change, anti-sex trafficking and other contemporary aspects of feminism and development. As with any long-term and cross-cultural collaboration, the participants have moved in and out of this project depending on life circumstances. Indeed, some of the authors have changed jobs, given birth and handled life-changing events that underscore the complex and intersecting dimensions of their personal and professional lives.

This book is designed to be a research-based interdisciplinary text and therefore is of particular interest to academic researchers and students in subjects such as Development Studies, International Relations, Feminist Studies and Geography. The content of many of the chapters is also useful to practitioners and policy makers given the authors' engagement with projects and concerns that relate directly to political and institutional aspects of urban planning, land reform, immigration policy and resource management. In addition to the breadth of topics and geographical locations, the essays in this volume provide a focused approach to the critical intersections of gender, development and globalization. Finally, the book is organized around themes and action items in gender, development and transnational feminism that are both relevant and empowering. Hopefully these essays will further motivate students, scholars and activists to be more globally informed, intellectually engaged and critically aware of social justice issues.

Ibipo Johnston-Anumonwo and Ann M. Oberhauser

Acknowledgments

This book was inspired by the African proverb that states: "It takes a village to raise a child." The support and efforts of our own academic and personal "villages" mentioned below have been instrumental at every stage of this project. First, we would like to thank the contributing authors for their willingness to meet deadlines, respond to reviewers' comments and provide assistance at a moment's notice. We greatly admire the dedication of these authors to the communities and the environments where they conduct research.

We also benefitted from the assistance and support of Faye Leerink, Geography Senior Editorial Assistant at Routledge, for professionally guiding us through the editing process. In addition, we are indebted to Erin Johns Speese for her efficient work editing the chapters and preparing the manuscript. Ram Alagan was also helpful in administering the Drop box folders for this project.

The following external reviewers of the book chapters provided constructive feedback: Fathima Ahmed, Nancy Akinyi Omolo, Debbie Budlender, Joe Curnow, Amarasiri de Silva, Clara Greed, Cecilia Green, Cynthia Gorman, Lori Hanson, Paddington Hodza, Julietta Hua, Jennifer Hyndman, Michele Masucci, Brent McCusker, Kate McLean, Janet Momsen, Barbara Penner, Geraldine Pratt, George Roedl, Michael Sheridan, Neera Singh, Jennifer Smith, Padmini Swaminatan, Charlotte Wrigley Asante, Stephen Young and Tiantian Zheng. We are also grateful to the anonymous reviewers of the book proposal for Routledge Press. In addition, Linda Peake and Monica Ogra gave timely and perceptive suggestions for the introductory chapter during a critical stage of this project. We would also like to thank colleagues at the 2012 Association of American Geographers conference sessions who generously shared their views on this research.

Several other colleagues and students shared their insights and work during this project, including Martina Caretta, Maureen Hays-Mitchell, Ragnhild Lund, Smita Mishra-Panda, Gina Porter, Alanna Markle, Bradley Wilson, Jennifer Rogalsky, Abra Sitler and Muriel Yeboah. Thanks to students in the feminist geography seminar who raised critical questions and provided a forum to discuss many of the issues in this book. Working with these accomplished scholars and students has been an enriching experience.

Throughout the project we gained from the intellectual feedback, goodwill and encouragement of colleagues such as Katherine Nashleannas, Seth Asumah, Iheanyi Osondu, Karen Culcasi, Agnes Musyoki and Nthaduleni Nethengwe. The mentoring and scholarly work of Susan Hanson, Cindy Katz, Janice Monk, Janet Momsen, Richard Peet, Daniel Weiner and John Willmer also helped us develop the academic tools to ask more inclusive research questions and to learn to be more aware of where our research is taking us. Institutional backing from West Virginia University's Geography Program and Center for Women's and Gender Studies, and SUNY Cortland's Geography Department and Office of Research and Sponsored Program is also appreciated.

Finally, this book would not have been possible without the ongoing support, patience and understanding of our families. Our parents (Peter and Sanny Oberhauser and Rhoda Johnston) inspire us to be compassionate and intellectually engaged with the world around us, our children (Benjamin, Frances, Obi and Kachi) give us important perspectives about life's possibilities and limitations, and our partner and brother (Howard and Akintunde) provide a sounding board and a hot meal at the end of the day. This book is dedicated to our "village" of family, friends, colleagues and students.

With all this assistance, we are responsible for whatever deficiencies remain.

Ann M. Oberhauser and Ibipo Johnston-Anumonwo

Introduction

Engaging feminism and development – worlds of inequality and change

Ibipo Johnston-Anumonwo and
Ann M. Oberhauser

> The movement for change is a changing movement, changing itself, de-masculinizing itself, de-Westernizing itself, becoming a critical mass that is saying in so many different voices, languages, gestures, actions: It must change; we ourselves can change it.
>
> (Rich 1984: 228)

In the early 1980s, Adrienne Rich challenged Western feminists to move beyond the social and economic homogeneity and privilege of Western feminism in order to confront global inequalities and attain positive social change. Thirty years later, this challenge still resonates with feminists who oppose persistent inequalities under neoliberal globalization and uneven development. Efforts to mobilize and empower women include transnational feminist movements that support women's economic livelihoods and access to resources, as well as policies to combat gender-based violence (Peake and de Souza 2010; Ferree and Tripp 2006). These and countless other movements that advocate for gender equality and social change make up the global "voices, languages ... and actions" that are part of Rich's call for change.

The chapters in this book draw from these movements to engage with both feminism and development across diverse global contexts. Following the introduction, the second part of this chapter examines feminist theories and practices that challenge neoliberal globalization and offer a critical lens to explore gendered socio-spatial processes at both local and global scales. The third part focuses on contributions of the book to our understanding of the ways in which women and communities in the global South and North are both impacted by and shape globalization, participatory feminist methodologies and the intersection of social and ecological factors in environmental change. An overview of individual chapters follows in the fourth part with special attention to analytical frameworks and praxis employed in these feminist studies of globalization and development. As a whole, this discussion highlights the collaborative research that is conducted by many of the authors, as well as the collective nature of producing this book with scholars and activists from diverse academic and geographic backgrounds.

Feminist approaches to global development

Beginning in the mid-twentieth century, feminist approaches to gender relations and international economic growth have informed the field of gender and development. Early analyses concentrated on women and their marginalization in the development process. Increasing attention to gender roles and power relations in global economic development shifted the focus to include issues such as women's empowerment and social movements (Sen and Grown 1987). The Chipko Movement and the Green Belt Movement, in India and Kenya respectively, are examples of environmental activism that drew attention to women's access to natural resources in ways that would enhance their livelihood strategies (Agarwal 1995; Rocheleau *et al.* 1996; Maathai 2006). Feminist perspectives on development are more balanced by integrating men's gendered experiences without losing the emphasis on women. Finally, recent views on economic globalization and development critically examine transnational relations across multiple landscapes of power and conflict and the intersection of feminist theories and practice, or praxis. (See Beneria (2003), Mohanty (2003) and Momsen (2010) for an overview of gender and development approaches.) These shifting discourses on feminism, globalization and development capture the socially-embedded nature of gender relations during different eras, as well as in different contexts of the global South and North.

The collection of essays in this book addresses the connection of scholarship and activism through uniting considerations that are central to both feminism and development. Feminism is broadly defined as a critical practice that engages with power relations and focuses on gender as a socially constructed category. These gendered power relations impact access to and control over material resources and shape ideological perspectives of gender identity (Alexander and Mohanty 1997). Furthermore, feminist scholarship presents theoretical frameworks, methodologies and empirical analyses of how women and the gendered relations within which they are situated affect the development process (Cornwall *et al.* 2007; Beneria 2003; Momsen 2010). These scholars examine connections among diverse geographical locations and related socio-economic processes within colonial and neocolonial contexts that influence gendered livelihoods, social movements and control over resources. Thus the intersection of feminism and development presents analytic opportunity for research studies to critically examine gender relations in the global South and North (Cornwall *et al.* 2007).

Recent feminist theories and practices examine neoliberal globalization, capitalism and patriarchy and "the multiple ways they (re)structure colonial and neocolonial relations of domination and subordination" (Nagar and Swarr 2010: 5). One of these theories, transnational feminism (TNF), has emerged from feminist and gender studies as a means of challenging traditional conventions and boundaries of nation-states and instead emphasizing fluid circuits of capital and labor at the global scale (Pratt 2012). This approach offers both theoretically- and empirically-based insights on power relations, inequality and the social construction of gender in a "world without borders" (Grewal and Kaplan 1994; Moghadam

2005; Mohanty 2003). TNF adds to critical perspectives on development and globalization by examining how women's lives are connected across various sets of transnational relations and practices and what can be done to improve women's lives through transnational feminist praxis.

Additionally, TNF has expanded our understanding of feminism, development and globalization by highlighting the intersection of social categories and spatial scales as well as emphasizing the role of struggle and activism in diverse contexts (Katz 2004; Swarr and Nagar 2010). Transnationalism examines political practices and economic livelihoods among individuals and communities in multiple geographic locations who are mobile and interconnected through fluid and contested boundaries (Grewal and Kaplan 1994; Ferree and Tripp 2006). Feminist scholars incorporate gendered aspects of these transnational spaces and processes that are useful in analyzing, for example, the flows and local impacts of migration (Mahlar and Pessar 2001; Pratt and Yeoh 2003; Silvey 2004). In sum, transnational feminism provides a critical analytical framework that sheds light on the intersections of gendered socio-spatial processes in local and global arenas. The chapters in this collection use feminist approaches in general, and elements of transnational feminism where applicable, to demonstrate many of these intersections.

A disciplinary approach to feminism and development that overlaps with and incorporates many aspects of transnational feminism is feminist geography. Scholars and activists in this field navigate multiple scales, both conceptually and empirically, in order to challenge conventional boundaries that situate structures and processes within local, regional, global and nation-state systems (Pratt and Yeoh 2003; Gibson-Graham 2006; Katz 2004). Drawing from this cross-border and cross-scalar stance, feminist geography engages with the intersection of alternative spaces and boundaries such as those that occur in and among communities, households and individuals. The mobile nature of both capital and people is evident through global processes that include investment by multinational firms, transnational trafficking of human beings and global flows of finance capital.

Feminist analyses of hierarchical power relations in development studies discourse also examine colonial and neocolonial dimensions of the global landscape that situate socio-economic and political realities in historical context (Beneria 2003; Mohanty 2003). Efforts to deconstruct these power relations include analyses of the intersection of race, gender, age, class and other axes of power that make up the often fluid identities of individuals and communities (Ferree and Tripp 2006). Nagar and Swarr (2010) show that this hierarchical framework must be challenged in order to develop transgressive and emancipatory interventions based on transnational solidarities and collaborations. Thus feminist and transnational approaches emphasize cross-border and dynamic relations that connect people and places and provide alternative views on often static binaries imposed by conventional approaches to globalization (Mohanty 2003).

Finally, feminist analyses of globalization and development are grounded in praxis and engagement with communities in ways that contest hegemonic socio-economic processes (Nagar 2006; Ferree and Tripp 2006). Praxis, or the integration of theory and practice, is useful in analyzing the socio-spatial relations of capital as it structures and restructures both material and discursive places and time periods. As Mohanty argues, feminist scholarship

> is a directly political and discursive practice in that it is purposeful and ideological. It is best seen as a mode of intervention into particular hegemonic discourses ...; it is a political praxis that counters and resists the totalizing imperative of age-old 'legitimate' and 'scientific' bodies of knowledge.
>
> (Mohanty 2003: 19)

Indeed, praxis is embedded in struggles for empowerment and advancement of marginalized groups as they overcome barriers and challenge forces that limit their access to and control over land, economic livelihoods, mobility and other vital resources. Feminism is grounded in theorizing inequality and power relations, as well as activism to expand opportunities and improve the status of under-represented and marginalized groups throughout the world (Ferree and Tripp 2006; Gibson-Graham 2006; Beneria 2003). The link between theory and activism in transnational feminism is the basis for its use of praxis as an analytical and practical approach (Nagar and Swarr 2010). An important dimension of this activism is collective engagement among and within communities that forms through social struggles and change to improve the economic self-determination and social status of women with "different voices, languages, ... and actions" (Rich 1984: 228). As exemplified in many of the chapters in this collection, these social movements can reinvent identities and power relations in progressive ways.

Gendered worlds of development and globalization

This book integrates development and globalization through a collection of critical feminist perspectives on the economy, public policy, the environment and societal structures that shape women's, and men's, gendered worlds. Case studies from communities and regions in South Asia, sub-Saharan Africa, North America and the Caribbean illustrate diverse, yet related socio-economic and political contexts in which women struggle to empower themselves in order to negotiate discursive and material barriers in their everyday lives. Contributors to this book draw from research projects that engage with the spatiality of gender relations, especially those that entail social and economic inequalities. These projects critique neoliberal development and explore circumstances that promote equal power relations in the socio-spatial arenas of global politics, the economy and the natural environment. The authors focus on how people negotiate gendered aspects of economic livelihoods and social change in the development process. Their research provides insight on the links between gender and economic strategies across local, regional and transnational scales. The following discussion addresses areas of emphasis in this book and specifically its theoretically- and empirically-informed investigations of neoliberal globalization, feminist methodologies and environmental change.

The first area of emphasis: neoliberal globalization, features prominently in cross-border interactions, livelihoods and control over resources discussed in the previous section. Several chapters in this collection address neoliberal globalization as it relates to policy and institutional aspects of gendered development. National and international policy measures that focus on privatization, reduced state support and free market forces are often couched within neoliberal capitalism. These policies affect migration, production, public services, land reform and other societal matters (Beneria 2003; Gibson-Graham 2006). Although gender analyses have become more common in many international development organizations, feminist perspectives such as those offered in this collection provide a different outlook to social and economic transformation.

In critiques of neoliberal globalization, feminist scholars examine the impact of colonial and neocolonial relations on economic livelihoods and distribution of natural resources (Johnston-Anumonwo and Oberhauser 2011). Neoliberal capitalism builds on inequality associated with colonial and neocolonial relations and, in some cases, increases gender disparity and exploitation through multinational investments, global trade and international labor mobility (Beneria 2003; Gibson-Graham 2006; Nagar *et al.* 2002). Transnational feminism includes both conceptual and methodological possibilities to break down conventional borders and operations of the nation-state that contribute to these gender disparities and marginalization of women. Some scholars claim, however, that this alternative notion of transnational spaces is complicit in "the reproduction of patriarchy beyond national borders" (Pratt and Yeoh 2003: 162). Some of the chapters in this collection take these aspects of TNF to critically analyze neoliberal institutions such as non-governmental organizations (NGOs), governments and international organizations.

The second area of emphasis in this book focuses on active engagement with subjects and participatory methodologies in the research process. In several chapters, feminist methodologies provide important ways to advance critical and participatory research and apply it to real world issues such as gendered responses to microcredit programs, climate change, natural resources management and land reform. Through their involvement with grass-roots organizations and development projects, these authors demonstrate how participatory techniques support the knowledge and action to enhance gender equity, sustainability and empowerment. Participatory research is widely practiced in many transnational and critical feminist projects in developing regions of the global North and South. However, many scholars note that this methodology is often embedded in hegemonic and ethnocentric social relations that reflect Western and feminist practices and production of knowledge (Cornwall *et al.* 2007; Alexander and Mohanty 1997: Swarr and Nagar 2010). In light of what Mohanty refers to as methodological universalism where Western research constructs a monolithic image of the "Third World Woman," she argues for "careful, historically specific generalizations responsive to complex realities" that may also form strategic political affinities (2006: 37).

In some cases, contemporary forms of feminist research utilize recent developments in participatory geospatial technologies that include Geographic Information Systems (GIS), gender mapping and Global Positioning Systems (GPS) to identify gender inequalities in divisions of labor and in decision-making within communities. In many cases, these techniques examine the needs of both male and female participants in matters such as land reform, access to water or disaster relief with input and guidance from researchers. When used appropriately and in light of feminist critiques of and insights on GIS and participatory research (Rocheleau and Edmunds 1997; Kwan 2002), these and other methods build capacity in households and communities as well as raise awareness of institutional barriers and opportunities.

The third area of emphasis, human–environmental relations, plays a key role in feminist inquiries of how natural resources are linked to economic strategies and social relations. For example, environmental change and the intersection of social and ecological factors are recognized as critical areas of concern in the rapidly growing fields of feminist political ecology and GED (Gender, Environment and Development) (Rocheleau *et al.* 1996). (See reviews of this literature by Resurrection and Elmhirst (2008), Elmhirst (2011) and Cruz-Torres and McElwee (2012)). This work examines how the use of, access to and control over natural resources and the environment are socially embedded. In this book, case studies from sub-Saharan Africa and South Asia represent innovative approaches to identifying how individuals and communities adapt to and mitigate climate change and how they cope with conservation challenges. These studies furnish important insights about how socially-inscribed divisions of labor interact with dynamic environmental systems, framing our understanding of gendered outcomes.

By engaging both theory and practice, this book contributes to the field of transnational feminism through gendered perspectives on global development that include cross-border and cross-scalar connections. Analytical frameworks or concepts are rarely clear-cut. Globalization processes are complex, and neoliberal policies have mixed effects, as exemplified in the studies in this collection on service provision, microfinance, diasporic flows or sex trafficking. These policies sometimes improve access and facilitate empowerment for some women (and men), but at other times exclude or leave women at an impasse because of ideological, political or socioeconomic hurdles. Indeed, the notion of empowerment is heavily contested in feminist practice, with much conflicting evidence in the global development literature. Women may not always be able to initiate transformations, or effect change, but there are some situations where they are able to improve their conditions and gain some level of autonomy in their everyday lives. One must therefore be cognizant of the particular circumstances under which engagement with development programs or grass-roots organizations can truly enhance the goals of gender equity, sustainability and empowerment.

Similarly, as noted above, one cannot adopt a one-sided celebratory tone to the promise of transnational feminism. Granted that changes associated with globalization have often opened up possibilities for women to express their views and concerns beyond local and nation-state boundaries (Ferree and Tripp 2006), however, channels of transnational discourses do not always empower women or make their contributions visible. Moreover, many scholars question the ability of feminist research to enhance gender equity through the frameworks and techniques that guide this approach. Yet, fundamental analytical insights of transnational feminism remain pertinent for articulating the interconnections across feminism, globalization and development.

In addition, the mixed methods in many of these studies allow more fully-informed, inclusive and pragmatic inquiries. Some explicate the efficacy of a gender lens in mitigating inequities in access to land, redressing time poverty, or alleviating burdens of natural disaster and displacement. Others show that the construction of gendered agency is relevant in feminist understandings of women's and men's roles in development processes, be they in political activism or environmental adaptation. Inherent in the research process are complicated relationships including the new hierarchies that feminist methods in general and participatory methods in particular can create.

Collectively, we attempt to employ critical feminist and development theories that engage with non-hierarchical and fluid power relations to reinforce the connections between geographically-situated relations of power and bring actors and struggles in alignment with each other. As stated by Badola *et al.* in Chapter 11, any development initiative "when done in a way that directly includes objectives to reduce and ultimately eliminate gender-based inequities as both the means and the end can support a broader and deeply meaningful transformative process in and around the world." This outlook on feminist geography presents a common disciplinary framework of working to understand and to improve human wellbeing through advancing gender equity in the global North and South. In particular, cross-cultural perspectives unveil ways to dismantle oppression on the basis of gender, race, ethnicity, sexuality and other social axes of difference by exploring how these measures and approaches empower individuals and communities and challenge patriarchal and hegemonic patterns of power.

Global perspectives on gender and space

The sections in this book are organized around feminist approaches to neoliberal globalization, research methodologies and environmental change. As mentioned above, these topics reflect aspects of contemporary feminist studies on issues that broadly include conceptual and empirically-based investigations of how gender relations play a role in women's ability to participate in sustainable livelihoods at the local and global scales. Analyses of gendered labor are central to feminist studies on globalization and development, and many chapters in the book address women's work in one form or another. For instance, authors in the first section of the book examine various aspects of women's work and livelihoods (ranging from micro-enterprise to sex work) and the everyday challenges of reproductive labor associated with access to urban water resources as well as gendered dimensions of macroeconomic policies and practices.

The second section focuses on feminist fieldwork and research methodologies, exploring ideas about participatory research and interconnections of the local and the global, as well as the global North and South. These authors highlight the importance of advocacy in feminist scholarship and the need for the kind of applied research that brings to light (rather than obscures) diversity and gender differences. The studies in this section report findings of case studies as well as analytical models for future studies that aspire to use a transnational feminist framework. The final section examines interconnections between gender, the environment and community-based development. The research accounts for local knowledge and institutional structures to yield fuller insight on how global systemic factors explain social vulnerability and propose pragmatic gender-sensitive steps for climate adaptation, disaster recovery and sustainable (ecological) development. Consistent with the other empirically-grounded analyses in this volume, this section reports on institutional and individual strategies undertaken in response to prevailing and interlinked global economic and environmental crises. The remaining part of this introductory chapter briefly describes the original research contained in the book's eleven chapters while discussing significant elements that connect chapters in each of the three sections.

Given contemporary realities of global restructuring, chapters in the first part of the book, *Feminist perspectives on neoliberal globalization*, examine conceptual topics and policies about provision of services and transnational migration flows. The authors in this section evaluate political and economic programs and policies on sanitation and financial services as well as transnational movements like diasporic exchange and sex trafficking. The particular services of urban sanitation and microcredit are good examples for evaluating whether alternative arrangements for meeting these common demands address women's concerns. As noted by Njeru *et al.* and Samarasinghe, access to safe drinking water and the elimination of poverty and gender-based violence are highlighted in global initiatives including the 2015 UN Millennium Development Goals and the UN Anti-Sex Trafficking Protocol. By examining government, non-government and other private initiatives that supply these services, the findings of these studies inform analysts about gendered consequences of neoliberal policies that are being implemented across the global South. For example, taking into consideration the spread of neoliberal-inspired economic reforms across the global South, Chapter 1 by Njeru *et al.* and Chapter 2 by Aladuwaka and Oberhauser examine public sanitation services and microcredit services, respectively, in recognition that hardships associated with access to these services are gendered. Using the case of the commercialization of public toilets in Nairobi, and the provision of microfinance programs in Sri Lanka, the studies find that effects of privatization of toilet services reveals mixed results about improvement in sanitation access for women's wellbeing; and the opportunity to get "out of the kitchen" and participate in income-generating work offers rural Sri Lankan women some degree of enhanced spatial mobility and household autonomy—a possibility enabled in large measure by microfinance programs.

The next two chapters in this section demonstrate clear links to transnational feminist research through explicit interrogations of policies pertaining to global interaction. In Chapter 3, Mullings analyzes another emerging neoliberal policy associated with the important influence of members of the diaspora in development circles. Especially in the past two decades, scholars and policy makers have drawn attention to the importance of diasporic exchange to the processes of social transformation in the global South. Few studies, however, have looked specifically at the ways that social constructions of gender maintain, complicate or unsettle political and social hegemonies in the places with which diaspora members maintain their ties. Mullings explores the entanglements between Jamaica's emerging state policy orientation and existing gender inequalities. And in the last chapter of this section (Chapter 4), Samarasinghe details the intricate web of networks that connect many nation-states across the globe to argue that transnational migration policies and divisive political ideologies on sex work are two main obstacles to global anti-sex trafficking policy initiatives. The chapter corroborates previous case studies of transnational sexual economies (Dewey 2008) but moves beyond regional coverage in order to emphasize the increasingly global nature of trafficking since the passage of the UN Anti-Sex Trafficking Protocol of 2000. A broad-based examination of the impasse in global anti-sex trafficking policies illustrates how female sex traffic victims are situated in the middle of ideological debates while powerful states use their influence to manipulate the policy initiatives of dependent states.

In sum, using national case studies and cross-national comparisons that highlight problems associated with gender inequity to scrutinize divergent perspectives and initiatives, the studies in the first section of the book add to the literature on the varied gendered impacts of particular economic and political processes in the neoliberal global economy. The research presented in these chapters covers standard areas of scholarly inquiry on interconnections of gender, globalization and development but raises questions about policy and institutions that maintain livelihoods, particularly in countries experiencing economic crises. One significant conclusion from the feminist inquiry of these multifaceted research questions is that consequences of neoliberal globalization are not uniform and should not be regarded as such.

The second part of the book, *Gendering the field: participatory feminist research*, features an array of methodologies used in feminist studies with an emphasis on praxis and participatory research. The four chapters in this section illustrate applications of these methodologies and practical lessons learnt from the field that could both inform future policy and empower women. Taken together, these chapters provide grounded, yet theoretically-informed, accounts about doing and engaging in feminist fieldwork.

This section starts with Chapter 5, in which Oberhauser explores themes related to gendered livelihoods, praxis and intercultural research in the global North and South. The methodological approach in the study emphasizes positionality and diversity among researchers and participants as well as the potential for transnational feminism to empower women and marginalized communities. The empirical analysis addresses socio-spatial relations among women's economic collectives in two regions of South Africa and the United States. Transnational feminism encompasses intercultural aspects that engage with multiple identities and power relations among researchers and participants. This analysis operates at various levels and draws from diverse social categories in order to develop inclusive and participatory feminist research. In light of the emphasis on working across borders, this chapter, like the preceding two, establishes a clear connection between gender, development and transnational feminism.

In Chapter 6, Bassa *et al.* analyze land use in peri-urban communities in South Africa while Naybor and Alagan study time use and mobility in rural communities in Uganda in Chapter 7. Both explore situations where gendered access to and control over resources stems from the legacy of systemic racism under apartheid and/or traditional patriarchal customs that discriminate against women. Mental maps, participatory-GIS, problem ranking exercises and Venn diagrams are used in focus group discussions in Durban to highlight the effectiveness of qualitative methods in examining men's and women's unequal access to land. And even though the use of participatory mapping has successfully involved female participants in defining their relationship to place, the study of women's time use and mobility in rural Uganda shows that with the recent advent of technology, there is the potential to fully collaborate with women in poor communities in the type of research that was formerly based solely on the observations of outsiders. A clear message from the two studies is that gender-sensitive qualitative methods, and participatory uses of information technology, give voice to otherwise marginalized women.

Complementing the clear connection to transnational feminist analysis in Oberhauser's chapter, Jaksch maps the complex relationships between Tanzania and South Africa, specifically concerning women's contributions to the liberation struggles in these countries. She argues that the rich visual rhetoric that once helped create identities and recognition for women political activists has now largely disappeared. Instead, representations of women's domestic work seem either to reinforce ideas about women's voiceless relegation to the private sphere or to suggest that this work is neither political nor related to liberation struggles. *The Virtual Freedom Trail Project* (VFTP) elucidates visual and textual portrayals of the struggle against colonialism and apartheid, demonstrating how the politics of women's struggles, as exemplified by Bibi Titi Mohamed, reflect agency, activism and grass-roots social justice movements that can transform gender roles and power relations.

Considering the environment as a key site of gender inequity, the third part of the book, *Gender, the environment and community-based development*, examines how environmental and natural disaster threats and recovery efforts are gendered. In the earth-environment tradition of geography, one of the main concerns about women's empowerment, especially in the global South, is access to natural resources such as land and water. Feminist geographers suggest that a gendered analysis is relevant in examining how to mitigate environmental and disaster impacts and to expand access to social and economic resources (Enarson and Morrow 1998; Rocheleau *et al.* 1996; Sultana 2010; Seager 2010). The chapters in this section of the book emphasize understudied topics in feminist political ecology using case studies from East Africa and South Asia. Combined, the analyses incorporate concepts about sustainable development and practices that include feminist ways of handling contemporary environmental and ecological challenges of climate change, natural disasters, eco-development and sustainable communities.

Pastoral livelihoods remain under-represented in the literature on gender and climate change. In Chapter 9, Wangui works toward filling this gap by analyzing how pastoralists in three Masai communities in southern Kenya construct climate change through their gendered livelihood experiences. She argues that social differences within pastoralist communities influence how climate change is constructed by different individuals and groups and stresses the importance of gender in shaping these constructions. Gender analyses indicate that men and women have shared experiences in the Masai communities since their responsibilities are relatively well defined. In Chapter 10, Alagan and Aladuwaka expound on a multifaceted methodology that is used to analyze the gendered dimensions of post-disaster rehabilitation and recovery following the 2004 Indian Ocean tsunami in Sri Lanka. Feminist political ecology and other related literature on disasters show that women tend to face greater barriers to recovery and are marginalized from the traditional top-down decision-making process (Sultana 2010; Enarson 2012). The research in this chapter underscores the importance of Participatory Gender Mapping along with oral narratives, ethnographic description and other qualitative methods in improving post-disaster rehabilitation and recovery efforts.

The final chapter, by Badola *et al.*, attempts to help redress the gender gap in the literature on biodiversity conservation in the global South by analyzing the gendered nature of eco-development and integrated conservation and development projects in protected areas of the Indian Himalayas. Because gender and class/caste-based hierarchies tend to collectively reinforce longstanding patterns of elite and male privilege and authority in rural India, this analysis expands our understanding of the conditions under which such projects can be a means to empower both women and their communities. The authors argue that by including objectives to reduce and ultimately eliminate gender-based inequities, eco-development can play a meaningful role in supporting broader transformative processes necessary to achieve sustainable and equitable development in and around the world's protected areas. This chapter, like many chapters in the volume, draws attention to the inclusion of men in feminist analyses of development and globalization processes and policies without losing the focus on women.

In sum, this book enriches our understanding of globalization and development through feminist analyses of economic livelihoods, public policy, resource management and social change. The chapters offer cross-cultural comparisons and case studies that add empirically-informed insights about these socio-economic processes and practices. For example, several authors relate their case studies to policy aspects of global development as exemplified by the role of national governments in global sex trafficking and land reform efforts in contemporary South Africa. This collection also provides innovative approaches to feminist methodology that include participatory methods and geospatial technologies. These research techniques are designed to engage with and empower participants to identify needs and implement solutions in their communities. Finally, studies in this book illustrate how the relationship between feminism and development traverses multiple scales through transnational struggles and grass-roots organizations, while emphasizing human rights and gender equity.

References

Agarwal, B. (1995) *A Field of One's Own: Gender and Land Rights in South Asia*, Cambridge, UK: Cambridge University Press.

Alexander, M. J. and Mohanty, C. T. (eds) (1997) *Feminist Genealogies, Colonial Legacies, Democratic Futures*, New York: Routledge.

Beneria, L. (2003) *Gender, Development and Globalization: Economics as if People Mattered*, New York: Routledge.

Cornwall, A., Harrison, R. and Whitehead, A. (eds) (2007) *Feminisms in Development: Contradictions, Contestations and Challenges*, London: Zed Books.

Cruz-Torres, M. L. and McElwee, P. (2012) *Gender and Sustainability: Lessons from Asia and Latin America*, Tucson, AZ: University of Arizona Press.

Dewey, S. (2008) *Hollowed Bodies: Institutional Responses to Sex Trafficking in Armenia, Bosnia and India*, New York: Kumarian Press.

Elmhirst, R. (2011) "Introducing new feminist political ecologies," *Geoforum* 42(2): 129–32.

Enarson, E. and Morrow, B. H. (eds) (1998) *The Gendered Terrain of Disaster: Through Women's Eyes*, Westport, CT: Praeger Publishers.

Enarson, E. (2012) *Women Confronting Natural Disaster: From Vulnerability to Resilience*, Boulder, CO: Lynne Rienner Publishers.

Ferree, M. M. and Tripp, A. M. (eds) (2006) *Global Feminism: Transnational Women's Activism, Organizing, and Human Rights*, New York: NYU Press.

Gibson-Graham, J. K. (2006) *The End of Capitalism (As We Knew It): A Feminist Critique of Political Economy*, Minneapolis, MN: University of Minnesota Press.

Grewal, I. and Kaplan, C. (eds) (1994) *Scattered Hegemonies: Postmodernity and Transnational Feminist Practices*, Minneapolis, MN: University of Minnesota Press.

Johnston-Anumonwo, I. and Oberhauser, A. (2011) "Globalization and gendered livelihoods in sub-Saharan Africa: introduction," *Singapore Journal of Tropical Geography* 32(1): 4–7.

Katz, C. (2004) *Growing Up Global: Economic Restructuring and Children's Everyday Lives*, Minneapolis, MN: University of Minnesota Press.

Kwan, M. P. (2002) "Feminist visualization: re-envisioning GIS as a method in feminist geographic research," *Annals of the Association of American Geographers* 92(4): 645–61.

Maathai, W. (2006) *The Green Belt Movement: Sharing the Approach and the Experience*, New York: Lantern Books.

Mahlar, S. J. and Pessar, P. R. (2001) "Gendered geographies of power: analyzing gender across transnational spaces," *Identities* 7(4): 441–59.

Moghadam, V. (2005) *Globalizing Women: Transnational Feminist Networks*, Baltimore, MD: Johns Hopkins Press.

Mohanty, C. T. (2003) *Feminism without Borders: Decolonizing Theory, Practicing Solidarity*, Chapel Hill, NC: Duke University Press.

Momsen, J. (2010) *Gender and Development*, 2nd edition, New York: Routledge.

Nagar, R. (2006) *Playing with Fire: Feminist Thought and Activism Through Seven Lives in India*, Minneapolis, MN: University of Minnesota Press.

Nagar, R., Lawson, V., McDowell, L. and Hanson, S. (2002) "Locating globalization: feminist (re)readings of the subjects and spaces of globalization," *Economic Geography* 78: 257–84.

Nagar, R. and Swarr, A. L. (2010) "Theorizing transnational feminist praxis," in A. L. Swarr and R. Nagar (eds) *Critical Transnational Feminist Praxis*, Albany, NY: SUNY Press, pp. 1–20.

Peake, L. and de Souza, K. (2010) "Feminist academic and activist praxis in service of the transnational," in A. L. Swarr and R. Nagar (eds) *Critical Transnational Feminist Praxis*, Albany, NY: SUNY Press, pp. 105–23.

Pratt, G. (2012) *Families Apart: Migrant Mothers and the Conflicts of Labor and Love*, Minneapolis, MN: University of Minnesota Press.

Pratt, G. and Yeoh, B. (2003) "Transnational (counter) topographies," *Gender, Place and Culture* 10(2): 159–66.

Rich, A. (1984) "Notes towards a politics of location," in A. Rich, *Blood, Bread and Poetry: Selected Prose 1979-1985*, London: Little Brown & Co, pp. 210–31.

Resurrection, B. P. and Elmhirst, R. (eds) (2008) *Gender and Natural Resource Management: Livelihoods, Mobility and Interventions*, London: Earthscan.

Rocheleau, D. Thomas-Slayter, B. and Wangari, E. (eds) (1996) *Feminist Political Ecology: Global Issues and Local Experiences*, New York: Routledge.

Rocheleau, D. and Edmunds, D. (1997) "Women, men and trees: gender, power and property in forest and agrarian landscapes," *World Development* 25(8): 1351–71.

Seager, J. (2010) "Gender and water: good rhetoric, but it doesn't 'count'," *Geoforum* 41: 1–3.

Sen, G. and Grown, C. (1987) *Development Crises and Alternative Visions: Third World Women's Perspectives*, London: Earthscan.

Silvey, R. (2004) "Power, difference and mobility: feminist advances in migration studies," *Progress in Human Geography* 28(4), 490–506.

Sultana, F. (2010) "Living in hazardous waterscapes: gendered vulnerabilities and experiences of floods and disasters," *Environmental Hazards* 9: 43–53.

Swarr, A. L. and Nagar, R. (eds) (2010) *Critical Transnational Feminist Praxis*, Albany, NY: SUNY Press.

Part I

Feminist perspectives on neoliberal globalization

1 Gender equity and commercialization of public toilet services in Nairobi, Kenya

Jeremia N. Njeru, Ibipo Johnston-Anumonwo and Samuel O. Owuor

Introduction

With the promotion of private-sector development since the 1980s, neoliberalism has become a dominant ideology underpinning the global political economy, providing "the context and direction for how humans affect and interact with non-human nature and with one another" (Heynen and Robbins 2005: 5). A key feature of neoliberalism in the development process is its increased focus on market-oriented reforms in the global South through the delivery of basic municipal services such as provision of water and sanitation by private-sector actors (Bakker 2013). This shift in management of basic municipal services from the public to the private sector (or running them as if they were not public goods) is linked to changing household gender relations. In fact, feminist critiques of neoliberalism note that when services that were previously provided by the state become privatized or commercialized, workload for procurement of services by women intensifies alongside other social reproduction and household care-giving duties (Braedley and Luxton 2010). For example, since women are primarily responsible for water procurement to meet family members' health and hygiene needs, privatization of water services complicates their ability to fulfill those needs (Kerr 2004: 16).

In this chapter, we present a combined feminist and urban political ecology inquiry about the impacts of neoliberalism on the provision of water and sanitation services. The analysis draws from literature on urban gender equity, sanitation and neoliberal reforms as a framework for our research study of municipal toilet services in Nairobi, Kenya. Precisely because "neoliberal policies, and practices are not uniform and the effects are dependent on context" (Braedley and Luxton 2010: 20), the need for more studies conducted in the global South arises as a balance to the current preponderance of scholarship on the impacts of neoliberalism in the neoliberal heartlands of North America and Europe (for exceptions, see Beneria (2003), Ferguson (2006), Myers (2005) and Yeboah (2006)). This chapter contributes to the literature on neoliberalism and the provision of social services by highlighting the consequences of neoliberal policies on the commercialization of public services, especially water and sanitation in the global South.

The growing literature on toilet studies offers appropriate analytical room to examine the complex connections between women's lives and access to water and sanitation services in poor countries. As McDonald and Ruiters (2005) suggest, neoliberalization of these services transformed the managerial ethos of water and sanitation service organizations, political relationships and the relation between the state and citizens. An unsurprising result therefore, is that access to water and sanitation services has become important in debates about the advantages and drawbacks of neoliberal globalization. The centrality of water and sanitation provision to the realization of the United Nations Millennium Development Goals (MDGs) by 2015 has invigorated these debates (McGranahan and Satterthwaite 2006). In this study, empirical evidence from Nairobi on private-sector participation in the provision of municipal toilet services reveals mixed results about the link between privatization of sanitation services and improvement in women's wellbeing. Thus, the study supports the position that effects of privatization or commercialization are not uniform and should not be regarded as such.

Gender, sanitation services and public toilets

An estimated 2.5 billion people, the majority of them in the less developed world, do not have access to sufficient sanitation facilities (UNICEF/WHO 2012). Halving the proportion of people in the world without sustainable access to basic sanitation by 2015 is part of the seventh goal of the UN Millennium Development Goals (MDGs). World Health Organization (WHO) and World Bank initiatives that investigate the social and economic impacts of inadequate access to hygienic sanitation facilities highlight the health and mortality costs associated with diseases of poor water quality, sanitation and hygiene; the economic cost of time spent procuring water; and reductions in educational achievement due to illness and girls' attendance rates at schools (WSP 2012). In the case of Kenya, a 2007 study involving a survey of 2,905 households in three major cities (Nairobi, Kisumu and Mombasa) found that 70 percent of households reported that women were the primary water collectors. The study further reported that in Nairobi, the poor pay almost six times as much for water than the non-poor (CRC 2007).

As sanitation gains more global attention, public or communal toilets have gained interest among scholars and development practitioners alike. This interest stems from the assumption that safe and clean toilets are vital to establishing healthy, equitable and dignified communities. In the global North, where sanitation infrastructures are highly developed, scholars examined public toilets in the contexts of sex segregation, sexual identity, and accessibility and design and indicated their ramifications for social justice, citizenship and inclusiveness (see for example, Gershenson and Penner 2009; Greed 2003). Feminists seeking to understand how the availability, design and lack of toilets impact the wellbeing of women and girls focus on the role of the urban built environment in examining women's lives and their position in society (Gershenson and Penner 2009; Greed 2003; Jewitt 2011). Therefore, as key components of the urban built environment, public toilets

are important and revealing sites for discussions of the construction and maintenance of gender, sexual identity and power relations in general. Public toilets shape everyday urban experience on both an individual and collective level through their provision, location, and design. For instance, public toilets not only inform a woman's ability to move comfortably through a city, but also define what her "needs" are perceived to be by those in power and how she is expected to conduct herself publicly.

(Gershenson and Penner 2009: 9–10)

Studies further show that women often prioritize urban amenities such as public toilets differently than men, placing importance on location or design (Pain 1991; Valentine 1992). Availability and quality of public toilets, therefore, appear to be more meaningful to women. Indeed, reforms in the global North range from calls for more safe toilets for females to more gender-sensitive designs. And, in the global South, underdeveloped and inadequate sanitation facilities and infrastructure are also associated with important gender differences (Brewster *et al.* 2006). As Kothari notes, while lack of sanitation facilities affects men and women, sanitation needs and demands differ with gender:

Women have particular needs and concerns of privacy, dignity and personal safety. The lack of sanitation facilities in the homes can force women and girls to use secluded places, which are often away from home, exposing them to the risk of sexual abuse; in other circumstances, girls are forced to defecate only at home and help their mothers to dispose of human and solid waste.

(Kothari 2003: 20)

Additionally, evidence points to a link between girls' access to public toilets and education in the developing world. Research indicates that the availability of toilets in schools can, by providing privacy and dignity, enable girls to get an education, particularly after they reach puberty (Brewster *et al.* 2006). Mitchell (2009) underscores the importance of location and quality of toilets to school-girls in her work on the geographies of danger and, specifically, young students' accounts of safe and unsafe spaces in their schools. Documentation provided by students serves as important evidence of fear and anxiety regarding public toilets expressed by schoolgirls in action-oriented fieldwork undertaken in Swaziland, South Africa and Rwanda. Students' photographs, drawings and narratives depict unsanitary or insecure toilets in isolated locations or in locations where darkness, bushes, roads or other neighboring land uses (e.g. barracks) are bothersome elements (Mitchell 2009). Similar safety concerns in boarding schools are aired by an adult early-career female teacher, who remarked that: "we have been trying to get the administration to build another block of toilets closer to the dormitory. It is not safe" (Mitchell 2009: 70). By and large, public toilets in Africa are grossly inadequate for females.

Consequently, scholars, policy makers and activists have called for improved sanitation facilities because the improvements could be more meaningful and impactful to women in urban and non-urban locations (Dodman 2009; Undie *et al.*

2006). Zachary Asher Mason, a Peace Corps volunteer in Mali, observed in his 2008 blog:

> A toilet for girls does not mean that everyone who uses it will necessarily be able to read and write and it certainly does not guarantee future employment– but it provides the infrastructure necessary so that girls can at least stay in school through adolescence. . . .

In sum, while accounts in the literature about public toilets emphasize the gendered nature of access to sanitation services, scholars are only beginning to substantially engage with the neoliberal discourse of providing sanitation services. Feminist and urban political ecological perspectives on neoliberalism are especially relevant for interrogating types of sanitation reforms (existing or potential) that are important and meaningful to men and women.

Neoliberalism and sanitation services in sub-Saharan Africa: a feminist and urban political ecology analysis

> (N)eoliberalism has had substantial implications for African cities. Neoliberalism is clearly a large part of what led the exasperated former mayor of Zanzibar to ask me the rhetorical questions with which I began the chapter: it seems that water, drainage, and garbage do, indeed have an ideology attached to them now.
>
> (Myers 2005: 6)

Feminist scholars are increasingly aware of the gendered outcomes of the practices and policies of the global spread of neoliberalism, including in sub-Saharan Africa. Although there is no single feminist approach to neoliberal restructuring, but instead multiple approaches with a varying mix of feminist perspectives and methodologies, many argue that neoliberal processes have negative effects on women (Beneria 2003; Johnston-Anumonwo and Doane 2011; Runyan and Marchand 2000: 225). In their discussion about ways in which the transfer of governmental responsibilities to the private sector affects people's daily lives, Braedley and Luxton (2010: 12) identify social reproduction as one of the gender regimes that anchor neoliberalism. Women undertake the bulk of the socialization and reproductive household labor that includes mundane, uncounted and unpaid work associated with activities such as cleaning, obtaining water and food, cooking and taking care of family members. Yet neoliberalism's commitment to reducing government expenditures for social reproduction activities undermines state initiatives that improve women's conditions (Braedley and Luxton 2010: 15).

A significant portion of feminist analyses about globalization stress the effects of privatization, analyzing the negative and positive consequences of neoliberalism on the basis of gender. Since government cut-backs impact social spending, privatization affects women, especially low-income women, by causing them to bear additional reproductive burdens, including the care of household members, the sick and the elderly. Global perspectives on gender-water geographies shed some light on the implications of neoliberal water policies (O'Reilly *et al.* 2009; Sultana and Loftus 2011). Many less developed countries receive loans and other forms of debt relief from the World Bank and the International Monetary Fund (IMF) on the condition that "inefficient," state-run enterprises such as hospitals, water and sanitation agencies become privatized. Yet, according to Kerr:

> when water goes private, it is usually sold and controlled by a small number of primarily American or European transnational corporations at a much higher cost than prior to privatization. Some argue, therefore, that if privatization of essential services continues at the current rate, the essence of life will become unaffordable to most women and men living in poverty.
>
> (Kerr 2004: 16)

As such, many scholars believe that practices linked with neoliberalism have inherent ramifications for how urban residents everywhere access municipal services (Heynen *et al.* 2006).

Urban political ecology (UPE) provides another critical lens through which scholars have sought to understand impacts of neoliberalization. Central to UPE is the notion that complex and interrelated economic, political and cultural processes are largely responsible for bringing about environmental transformation in cities (Heynen *et al.* 2006). This approach recognizes that urban environmental transformations are not independent of gender, class and ethnicity (Swyngedouw and Heynen 2003). Considering the context of urban environmental outcomes of neoliberalism, feminist analyses can benefit from "the integrative, multi-dimensional language" of urban political ecology research (Myers 2005: 13–14).

UPE research also examines the privatization of water (Bakker 2007), highlighting the impacts of neoliberalism on urban sanitation services in the global South where, in recent decades, many governments have initiated restructuring in the water and sanitation sector (Smith 2004). The justification for this restructuring is the belief that, through introduction of market principles, greater efficiency and customer service will improve and extend services to all, including low-income groups. Yet, as is true in most of the global South, evidence in sub-Saharan Africa suggests that the shift from government intervention in the delivery of water and sanitation services to reliance on market forces and principles has not improved service access for many urban residents, especially the poor (McDonald and Ruiters 2005). For example, in South Africa, structural inequalities in toilet provision persist amidst efforts in sanitation reforms (Penner 2010). In fact, some scholars and civil society groups believe that these reforms may slow the progress of the seventh goal of MDGs in sub-Saharan Africa (Easterly 2009). Specifically, McKinley argues that unless subsidized by the state, commercialized (or privatized) public utilities rarely ensure broad access to poor households that find it difficult to pay commercial tariffs (2008).

Furthermore, under current neoliberal reforms, sanitation services have drawn little attention while water supply has attracted the majority of private sector investment (van der Hoek *et al.* 2010). According to Budds and MacGranahan, the lack of focus on sanitation "reflects the fact that while the private benefits from water are usually sufficient to create a considerable demand for water, the same cannot be said of sanitation" since users "are less willing to pay for safe sanitation, yet its provision is highly desirable from a public health perspective" (2003: 96). The expectation that households pay for an on-site (or on-plot) sanitation infrastructure complicates private investment in sanitation services. Those living in slum areas or informal settlements are underserved by the existing sanitation infrastructures; they lack the financial capacity to invest in on-site sanitation facilities and the security of tenure for their homes. Therefore, the provision of water and sanitation by private investors causes the marginalization of poor households. Indeed, intersectional feminist analyses recognize approaches that emphasize the intricate links across social categories, including gender, race, ethnicity, age and class. A combined feminist-urban political ecology perspective should be helpful in understanding the gender impacts of the ongoing neoliberal-inspired reforms of municipal toilet services in Nairobi.

Commercialization of municipal toilet services and Ikotoilet "toilet malls" in Nairobi

> Some years ago, answering a call of nature in Nairobi's central business district was a nightmare. All City Council toilets were run and managed by street boys and goons. People were always mugged or forced to part with more money by the street boys who were armed with human waste.
>
> (*Daily Nation*, June 17, 2009)

Until recently, municipal toilets in Nairobi have been managed by the Nairobi City Council (NCC), which owned and operated 138 on-street public toilets—most of which were developed during the colonial period (WSP 2004). In keeping with colonial sanitation planning, which was designed to mainly benefit the colonial elite, the toilets were developed to serve city residents and visitors (who were mostly Europeans) in commercial areas. The post-independence policy makers in Nairobi retained the colonial sanitation policy, which promoted the development of municipal toilets in commercial areas and excluded residential areas where households were (and are) expected to provide their own on-site sanitation facilities. Yet, for many residents in the city's poor areas, who cannot afford to install a private toilet in their residences, lack of municipal toilet development in residential areas has historically left them with little choice but to engage in unsafe sanitary practices such as open-defecation (Ngugi and Ndegwa 1992).

Beginning in the 1980s, a combination of neoliberal budgetary cuts and poor planning resulted in significant maintenance and management problems of existing toilet facilities (as the quote at the beginning of this section demonstrates). Most toilets fell into disrepair, became chronically unhygienic and barely functional. While the disrepair limited the ability of all Nairobi residents/visitors to use the toilets, evidence from a study (from which this chapter draws and discussed below in the gender analysis section) suggests that the lack of usable toilet facilities disproportionately affects women. As a female respondent in this study remarked,

> it is very difficult for them [that is, women] to dispose their waste unlike men who only unzip their trousers and relieve themselves in open areas.
> (Female respondent at Accra Road Facility, authors' parentheses)

This comment refers to the behavior of men in the city who, unable to access usable toilets, urinate or defecate in open areas such as alleys and green-spaces. Also, during the 1980s and 1990s, women became easy targets for street boys, who used the dysfunctional facilities as hideouts from where they extorted money from unsuspecting toilet users and passersby, threatening to smear them with human waste.

Since 2003, however, public toilet services in the city have significantly improved. Legislation that the Kenyan parliament passed in 2002, proposing radical policy changes in the water and sanitation sector, provided the impetus for this improvement. The legislation signaled a move towards privatization and a redefinition of the government's role in the management of water and sanitation (Wambua 2004). As a result of this legislation, municipalities across the country have transformed their water and sewerage departments into autonomous business units, whose operations have been modeled after private sector business practices.

In keeping with the vision of this legislation, the Nairobi City Council (NCC) adopted neoliberal policy recommendations of the Water and Sanitation Program-Africa (WSP-Africa) to commercialize its toilet services (WSP 2004). The WSP-Africa, a program of the World Bank, recommended that the NCC allow private operators to take full control of existing facilities and/or build, finance and operate new facilities on a contract basis. In accordance with the recommendations, the NCC entered into a partnership with the Nairobi Central Business District Association (NCBDA) to rehabilitate toilets located within the city center and transfer their operations to private individuals/entities. Given the large demand for public toilet services, developing new facilities along the lines of the WSP-Africa's build-finance-operate framework became pertinent. On October 30, 2007, the NCC placed an advertisement in a major local newspaper, *The East African Standard*, inviting applications from community organizations, NGOs and companies wishing to construct and manage new public toilets in Nairobi. On June 1, 2008, the NCC accepted an application from Ecotact Ltd, and signed a memorandum allowing the company to develop and manage new public toilets on public land in the city. According to the agreement, the company would initially build fifty facilities, branded "Ikotoilets," and operate them under a five-year, Build-Operate-Transfer arrangement. While no official explanation was given for why the application from Ecotact Ltd was the only successful one, the Kenya Anticorruption Authority approved it.

David Kuria, an architect by training, founded Ecotact Ltd in 2006 as a social entrepreneurship venture. Before founding the company, Kuria worked with the NCC and various NGOs in Kenya's sanitation sector to gain insights into the scale of Nairobi's sanitation problem. To Kuria, a solution to the problem required a new model of sanitation service delivery that would give public toilets an image makeover. The term "Iko" implies ecological in English and translates to "it's here" or "there is" in Kiswahili, Kenya's national language. Thus, "Ikotoilet" sends a message to city residents and visitors that a toilet facility is nearby. It also suggests a desirable, safe and hygienic place that serves an economic, social and environmental purpose. In fact, the Ikotoilet's slogan, "thinking beyond a toilet," is an attempt by Ecotact to remove the stigma from sanitation and make the toilet a desirable purchase for its customers.

Ecotact has developed fourteen multi-use facilities in Nairobi, which many Kenyans refer to as "toilet malls" (Figure 1.1). Each Ikotoilet consists of eight toilet/shower combination stalls, divided evenly among male and female units, with three to four additional urinals in the male units. These facilities also have kiosks, which, depending on the location, offer mobile money services and sell merchandise such as snacks, mobile phone airtime and beverages (Figure 1.2).

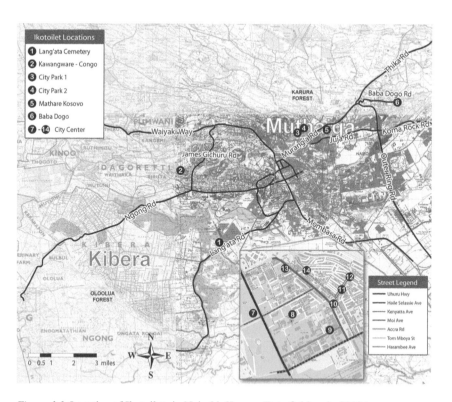

Figure 1.1 Location of Ikotoilets in Nairobi, Kenya. (Data fieldwork, 2012.)

Figure 1.2 Aga Khan Walk Ikotoilet. (Photo by Samuel Owuor, 2012.)

Additionally, Ikotoilet facilities located in commercial areas have outside veran-das for use by shoe-shiners and also offer opportunities for outdoor advertising. Ecotact leases kiosk and veranda spaces to third-party vendors. Of the fourteen facilities, three are in low-income areas of Kawangware-Congo, Mathare-Kosovo, and Mathare-Baba Dogo (Figure 1.3).

Ecotact's expectation is that the diversified revenue model that Ikotoilet toi-let malls represent ensures the sustainability and equitability of toilet services. The diversified revenue structure promises to keep user-fees relatively low and makes cross-subsidization of toilets developed in low-income areas possible. Indeed, about 17 percent of annual revenue from each facility is from advertis-ing, rent for the kiosk and shoe-shine space (Hussain 2011). Customers pay 10Ksh (US$0.13) for each toilet use and depending on the location, 20–50Ksh (US$0.25–0.63) to bathe. In low-income areas, residents pay subsidized rates of either 5Ksh (US$0.06) for single use or 100Ksh (US$1.25) for a monthly family package. The company has also customized Ikotoilet facilities to meet the all-around sanitation needs of residents—most of whom have no toilet or running water in their homes. The facilities are "one-stop-shopping" places, where residents can deposit their human waste, bathe and purchase safe drinking water.

Viewed as an innovative model of sanitation service delivery in poor countries, the Ikotoilet toilet malls have earned Kuria/Ecotact several awards, including Africa Social Enterprise of the Year 2009 at the World Economic Forum, the 2010 Public Service Award from the African Ministers' Council on Water and special recognition from the 2009 Clinton Global Initiative. It is important to note that the engagement of Ecotact in the management of public toilets in Nairobi reflects a shift in neoliberalism with respect to the privatization of public services in poor countries. Scholars and practitioners increasingly realize that neoliberal policies of privatization, involv-ing outright transfer of public functions and services to private entities, have not

Figure 1.3 Kawangware-Congo Ikotoilet. (Photo by Samuel Owuor, 2012.)

improved efficiency, quality and equal access to public services in the developing world as most development practitioners in major international organizations had assumed (McKinley 2008; Miraftab 2004). In light of this, many proponents of privatization, including the World Bank, the IMF and bilateral development agencies, such as the United States Agency for International Development and Britain's Department for International Development, now advocate public-private partnerships as the solution to inadequate provisions for public services in the fast-growing cities of the developing world (McDonald and Ruiters 2005).

Advocacy for public-private partnerships has coincided with interest among neoliberals about the role of social entrepreneurship in the provision of public services (Hervieux *et al.* 2010). Social entrepreneurship, which is viewed as a vehicle to restructure welfare, "involves building social partnerships between the public, social and business sectors" (Cook *et al.* 2003: 57), "while harnessing market behavior in the interest of public goals" (Dey 2010: 8). There are concerns, however, that more often than not, such partnerships serve as a means for "advancing the interests of the private sector and the market under the banner of sharing power with the poor and the state" (Miraftab 2004: 89). Overall, Kuria's Ecotact initiative has valuable social benefits, but given the concerns about differential impacts of neoliberalism and associated processes of privatization and public-private partnerships on the poor (women and men), a gender analysis of this commercial model of sanitation provision is necessary.

A gender analysis of Ikotoilet users

A gender analysis of commercialized public toilet services in Nairobi, and specifically their impact on women and men's living conditions, raises questions about who benefits, who loses and who is left out in the neoliberalized toilet malls. Are the facilities designed, geographically positioned, maintained and regulated in a manner that is gender sensitive and provides safe spaces for women (see for example, Mitchell 2009; Braedley and Luxton 2010: 12)? In this section, we draw from empirical research conducted in Nairobi to highlight gender differences in the use and perceptions of Ikotoilets and the commercialization of municipal toilet services as a whole.

During the month of June 2012, 126 Ikotoilet users, comprising an equal number of men and women, were surveyed about these issues. Systematic random sampling was used to select eighteen respondents at each of seven Ikotoilet facilities chosen for the field study. Questionnaires were administered to Ikotoilet users at three facilities located within the city center (Aga Khan Walk, Accra Road and Uhuru Park), two facilities in suburban areas (City Park and Langata Cemetery) and two low-income/slum areas (Kawangware-Congo and Mathare-Kosovo). In Figure 1.1, the locations of Aga Khan Walk and Accra Road facilities are indicated by numbers 10 and 11 respectively. Three female and three male research assistants administered the survey questions. The six research assistants also took relevant field notes, including photographs of the toilet facilities.

Overall, 73 percent of all surveyed users responded that they were satisfied with Ikotoilet services. The satisfaction level is similar between men and women (48 percent and 52 percent respectively). In slum areas, 60 percent of users were satisfied with Ikotoilets. These results echo the opinions of many respondents who were pleased with the benefits and quality of Ikotoilet services, as the following sentiments attest:

> They have helped many people, especially in the slums.
>
> (Man at Mathare Ikotoilet)

> I have a place to bathe and use a toilet unlike if there was no Ikotoilet.
>
> (Woman at Mathare-Kosovo Ikotoilet)

> They have water and you are provided with tissue paper.
>
> (Woman at City Park Ikotoilet)

> They are always clean and their services are good.
>
> (Woman at Aga Khan Walk Ikotoilet)

In contrast, the cost of these services to users revealed some negative attitudes about Ikotoilet facilities. Fifty percent of users felt that the cost of using Ikotoilet facilities was unaffordable. In slum areas, slightly more than half the respondents (53 percent) thought that the cost of using the facilities was unaffordable, with an even distribution between men and women. Opinions about affordability were not markedly different between men and women. Moreover, the majority (69 percent) of respondents claimed that women are the main beneficiaries of Ikotoilets; 70 percent of women compared to 55 percent of men felt that women benefitted most. Respondents who claimed that women benefitted most from Ikotoilets provided three explanations. First, the toilets address the privacy needs of women. A female respondent at the Accra Road facility stated: "Women cannot relieve themselves at any point just like men." In fact, 73 percent of women and 75 percent men in the overall sample thought that Ikotoilets offered sufficient privacy for all. Second, respondents thought that Ikotoilets assured the safety of all users, including women; 79 percent of respondents (80 percent of women) felt that the facilities offered sufficient safety. Third, the availability of Ikotoilet facilities and the services that are provided are especially valuable to women because of their unique physical and social needs. The notions that "women go often" and that "women need to change sanitary pads" were often brought up by both women and men in explaining why respondents felt women gained most from Ikotoilets. A woman at the Aga Khan facility and a man at the Uhuru Park facility, who responded that the toilets benefitted women more than men, gave these explanations respectively: "Due to the (women's) need to dispose pads and the Ikotoilets offer the services", and "Women . . . they have loose bladders and also their monthly periods and need to change pads."

In addition to lavatory needs, a key dimension of the Ikotoilet model is that it offers other uses besides human waste disposal. The study reveals that for those in slum areas, bathing is among the most popular uses of Ikotoilets. Likewise, for those in slum areas, Ikotoilets are an important source of drinking water. In particular, female patrons in these areas listed water as second to waste disposal in the importance of services that Ikotoilets provide. This finding corroborates those of previous studies in Nairobi's low-income areas, which highlight frustration about unmet water needs among inhabitants (Undie *et al.* 2006; CRC 2007).

In spite of the substantial positive ratings accorded to Ikotoilets, one area of inadequacy is clear. There is an overwhelming sense that the fourteen facilities developed thus far do not meet the demand for public toilet services in Nairobi. Eighty-one percent of respondents (79 percent of women and 84 percent of men) thought that the number of Ikotoilets was inadequate. Moreover, a major disappointment among many users of Ikotoilets was that the hours of operation were not sufficient. Whereas 62 percent of all surveyed users felt that the Ikotoilets' hours of operation (a daily average of fourteen hours) were sufficient, two-thirds of respondents in low-income areas felt that hours were *not sufficient* enough. Much lower percentages of users in the slums thought that the hours were adequate (only 41 percent of women and 35 percent of men)—a finding that we interpret as confirming greater expressed need for sanitation services in low-income residential areas. Moreover, the Citizens Report Card on satisfaction, dissatisfaction and inequities among social groups in urban Kenya concluded that "people want more public toilets" (CRC 2007: 36–7). That study, however, did not incorporate a gender analysis to examine women's and men's experiences or needs.

In terms of availability, a somewhat negative aspect for women was the time spent waiting to use Ikotoilets in comparison to men. While 82 percent of male users stated that it took them less than five minutes to attend to their sanitation needs, only 73 percent of women said it took them under five minutes. Furthermore, toilet malls do not provide sufficient numbers of stalls for women. As mentioned above, Ikotoilets have the same number of toilet stalls for both genders, but women are disadvantaged because the facilities have urinals for males, with no matching stalls for females (Figure 1.4). These findings support general results of studies on public toilet use, which show that women spend more than twice as long as men in these facilities and thus need more facilities than men. Indeed, Greed argues that "women should be provided with not just 'equal' provision but ideally twice as many facilities" (2003: 8).

Finally, a substantial number of Ikotoilet users agreed that the private sector should be responsible for the provision and management of public toilets. Sixty-two percent of respondents, evenly divided among men and women, preferred private-sector to public-sector provision of this service. Respondents in low-income areas were evenly divided between support for private and public provision of toilet services. Also, support for privatized public toilet management was nearly even between men (61 percent) and women (59 percent) living in the slum areas. Among both men and women, the reasons most commonly offered for supporting the private sector were sustained toilet cleanliness and the accessibility of toilet facilities. Women, however, also mentioned that these private-sector-operated toilets provided better privacy and safety. According to a user at the City Park facility, "If the public sector involves itself, then they will be closed due to poor standards of hygiene." In keeping with this broad support for private-sector provision, 65 percent of respondents thought that the privatization of public toilet operations in Nairobi city was a good idea (70 percent of women and 66 percent of men). But, in the slum areas, the viewpoint on this issue was evenly split, as only half of the respondents, 50 percent (of both males and females), believed that this was a good idea.

Although the survey implies a preference for private versus public provision of services, the Ecotact model is one in a range of options for private-sector provision

Figure 1.4 Inside the male compartment of the Aga Khan Walk Facility. (Photo by Samuel Owuor, 2012.)

of sanitation and public toilets that lie somewhere between for-profit multinational corporations and government provision. As a small-scale business initiative that is providing social benefits (including jobs and income to Kenyans) in addition to sanitation and water services, Ecotact represents a shift in neoliberal discourse that emphasizes social entrepreneurship and public-private partnerships as an alternative to the extreme of privatization. Yet, at this juncture (given the concerns that even with this shift in emphasis, private-sector interests and profit-seeking practices remain dominant) analysts may have to be tentative rather than conclusive about the social benefits of Ecotact. The summary of research findings in the concluding discussion of this chapter underscores the complicated and nuanced impacts of private and commercial initiatives.

Conclusion

Contrary to most feminist and urban political ecology analyses of the impacts of neoliberalism in sub-Saharan Africa, the ongoing privatization of basic municipal services in the region has not always led to increased socio-environmental inequities. Ikotoilet toilet malls, as one manifestation of the privatized provision of sanitation services, seem to have improved women's access to hygienic public toilet services in Nairobi. In this study, exceeding complaints about affordability, the main criticism of the toilet malls was that there are too few. Therefore, these types of reforms in water, sanitation and hygiene services, when developed and implemented in ways that take consideration of the local social, economic, cultural and environmental conditions, are likely to be more meaningful and helpful to women.

One of the findings from this study addresses spatial disparity among beneficiaries of these public-private investments and supports feminist and political ecology studies which show that transformations of urban (built and physical) environments under neoliberalism reflect (and accentuate) existing gender and social inequalities. For example, Penner (2010) draws attention to structural inequities and uneven access to toilets among urban residents. Bakker (2013) also refers to private sector investment in sanitation taking place in capital cities, and in more profitable locations within these cities, at the expense of toilet investments in rural areas and in low-income or slum neighborhoods, where the need for sanitation and water is greater.

Findings from this Nairobi study of private provision of municipal toilet services also confirm that socio-spatial inequities are evident at the intra-urban level. Clearly, users in slum locations do not show the same level of enthusiasm about toilet malls as those in non-slum neighborhoods. Specifically, compared to the overall sample, slum residents (apart from ranking access to water higher) also express lower satisfaction, report hours of operation as inadequate and generally convey less support for the private provision of public toilets. Given the greater number of toilet malls in the Nairobi Central Business District, the results of the research confirm a spatially uneven pattern of private-sector activity and toilet provision, especially in higher-profit, downtown locations, but less so in suburban areas. The disproportionate number of Ikotoilets in downtown Nairobi supports the above critique of privatized provision of public services under public-private partnerships, even when such partnerships involve social enterprises like Ecotact. Furthermore, by providing the same number of toilet stalls for men and women in each of its facilities, the practices of Ecotact Ltd validate the above feminist critique of the under-provision of public toilets.

Critics of neoliberalism claim that the commercialization of public services more often than not compounds the marginalization of the poor and of women. In the context of much of urban Africa, with minimal availability of public toilets, provision of these toilets is better than nothing. Thus, availability of public toilet services via neoliberal commercialization represents a step in the right direction. Payment for services that should ordinarily be provided free, subsidized and/or at affordable rates by the government shifts the cost to families and individuals. As scholars have shown, women disproportionately bear this cost of public service provision (Braedley and Luxton 2010; Kerr 2004).

The provision of commercial public toilets by private entrepreneurs can be seen as an initiative that offers both a service and more choices, but the choice to buy sanitation services is contingent on the ability to pay. Access to this paid form of toilet service is a situation of constrained choice, with very clear class and gender differentials. Political and economic policies that influence women's spatial mobility or access to urban spaces and sites can offer both "enabling and constraining possibilities" (Laws 1997: 62). Future studies are needed to examine the extent to which the case of commercial toilets in Nairobi illustrates contradictory processes of inclusion and exclusion.

In sum, the provision of urban services illustrates how feminism and urban political ecology offer important analytical entry points for inquiries on improved socio-environmental equity. In addition to examining the neoliberal policy of the commercialization of public toilets, this research provides an examination of gendered encounters within the urban built environment. Accordingly, the study views Ikotoilets as part of a changing urban built environment in Nairobi resulting from policies and practices of a global neoliberal ideology. In this sense, the approach reflects a focal point of feminist scholarship by interrogating how decisions that are made at international- and national-level seats of power affect women (and men) in their everyday lives.

References

Bakker, K. (2007) "The 'commons' versus the 'commodity': alter-globalization, anti-privatization, and the human right to water in the global South," *Antipode* 39(3): 430–55.

Bakker, K. (2013) "Neoliberal versus postneoliberal water: geographies of privatization and resistance," *Annals of the Association of American Geographers* 103(2): 253–60.

Beneria, L. (2003) *Gender, Development and Globalization: Economics as if People Mattered*, New York: Routledge.

Braedley, S. and Luxton, M. (eds) (2010) *Neoliberalism and Everyday Life*, Montréal and Kingston: McGill-Queen's University Press.

Brewster, M. M., Herrmann, T. M., Bleisch, B. and Pearl, R. (2006) *A Gender Perspective on Water Resources and Sanitation*. Available online at: www.unwater.org/downloads/bground_2.pdf (accessed May 16, 2013).

Budds J. and McGranahan, J. (2003) "Are the debates on water privatization missing the point?: experiences from Africa, Asia and Latin America," *Environment and Urbanization* 15(2): 87–114.

Cook, B., Dodds, C. and Mitchell, W. (2003) "Social entrepreneurship: false premises and dangerous forebodings," *Australian Journal of Social Issues* 38: 57–71.

CRC (2007) "Citizens' Report Card on urban water, sanitation and solid waste services in Kenya. Summary of results from Nairobi, Kisumu and Mombasa." Available online at: www.wsp.org/sites/wsp.org/files/publications/ 71220-0745708_Citizens_Report_ Card_ Summary_Kenya.pdf (accessed May 16, 2013).

Daily Nation (2009) "How Kenyan scooped global award," *Daily Nation*. Available online at: www.nation.co.ke/News/-/1056/612288/-/ukayxg/-/index.html (accessed May 16, 2013).

Dey, P. (2010) "The symbolic violence of 'social entrepreneurship': language, power and the question of the social (subject)," paper presented at The Third Research Colloquium on Social Entrepreneurship, Saïd Business School, University of Oxford.

Dodman, D. (2009) "Globalization, tourism and local living conditions on Jamaica's north coast," *Singapore Journal of Tropical Geography* 30(2): 204–19.

Easterly, W. (2009) "How the Millennium Development Goals are unfair to Africa," *World Development* 37(1): 26–35.

Ferguson, J. (2006) *Global Shadows: Africa in The Neoliberal World Order*, Durham, NC: Duke University Press.

Gershenson, O. and Penner, B. (eds) (2009) *Ladies and Gents: Public Toilets and Gender*, Philadelphia, PA: Temple University Press.

Greed, C. (2003) *Public Toilets: Inclusive Urban Design*, Oxford: Architectural Press.

Hervieux, C., Gedajlovic, E. and Marie-France, B. T. (2010) "The legitimization of social Entrepreneurship," *Journal of Enterprising Communities* 4(1): 37–67.

Heynen, N. and Robbins, P. (2005) "The neoliberalization of nature: governance, privatization, enclosure and valuation," *Capitalism Nature Socialism* 16(1): 5–8.

Heynen, N., Kaika, M. and Swyngedouw, E. (eds) (2006) *In the Nature of Cities: Urban Political Ecology and the Politics of Urban Metabolism*, London and New York: Routledge.

Hussain, K. (2011) "Providing Adequate Sanitation to the Base-of-the-Pyramid in Kenya: a report by Khuram Hussain, Acumen Fellow at Ecotact in 2011." Available online at: www.johnson.cornell.edu/Portals/0/PDFs/2nd%20place.pdf (accessed May 24, 2013).

Jewitt, S. (2011) "Geographies of shit: spatial and temporal variations in attitudes towards human waste," *Progress in Human Geography* 35: 608–26.

Johnston-Anumonwo, I. and Doane, D. (2011) "Globalization, economic crisis and Africa's informal economy women workers," *Singapore Journal of Tropical Geography* 32(1): 8–21.

Kerr, J. (2004) "From 'Opposing' to 'Proposing': finding proactive global strategies for feminist futures," in J. Kerr, E. Sprenger and A. Symington (eds) *The Future of Women's Rights: Global Visions and Strategies*, London: ZED Books, pp. 14–37.

Kothari, M. (2003) "Privatising human rights – the impact of globalization on adequate housing, water and sanitation. United Nations Public Administration Network Social Watch Report." Available online at: http://unpan1.un.org/intradoc/groups/public/documents/APCITY/UNPAN010131.pdf (accessed May 16, 2013).

Laws, G. (1997) "Women's life courses, spatial mobility, and state policies," in J. P. Jones, H. J. Nast and S. M. Roberts (eds) *Thresholds in Feminist Geography: Difference, Methodology, and Representation*, Lanham, MD: Rowman & Littlefield Publishers, pp. 47–64.

McDonald, D. A. and Ruiters, G. (eds) (2005) *The Age of Commodity: Water Privatization in Southern Africa*, London, UK and Sterling, VA: Earthscan.

McGranahan, G. and Satterthwaite, D. (2006) "Governance and getting the private sector to provide better water and sanitation services to the urban poor," International Institute for Environment and Development-Human Settlements Discussion Paper Series. Available online at: http://pubs.iied.org/pdfs/10528IIED.pdf (accessed May 16, 2013).

McKinley, T. (2008) "Foreword: whither the privatization experiment?," in K. Bayliss and B. Fine (eds) *Privatization and Alternative Public Sector Reform in Sub-Saharan Africa*, New York: Palgrave Macmillan, pp. xviii–xx.

Mason, A. Z. (2008) "On second wave feminism and toilets," a blog by a Peace Corps Volunteer in Mali. Available online at: http://zacstravaganza.blogspot.com/2008/11/on-second-wave-feminism-and-toilets.html (accessed May 16, 2013).

Miraftab, F. (2004) "Public-private partnerships: the Trojan horse of neoliberal development?," *Journal of Planning Education and Research* 24(1): 89–101.

Mitchell, C. (2009) "Geographies of danger: school toilets in Sub-Saharan Africa," in O. Gershenson and B. Penner (eds) *Ladies and Gents: Public Toilets and Gender*, Philadelphia, PA: Temple University Press, pp. 62–74.

Myers, G. A. (2005) *Disposable Cities: Garbage, Governance and Sustainable Development in Urban Africa*, Hampshire, UK and Burlington, VT: Ashgate.

Ngugi, G. and Ndegwa, G. (1992) "The status of sanitation: provision and use of public conveniences in the city of Nairobi, Kenya," *African Urban Quarterly* 7(1–2): 99–102.

O'Reilly, K. O., Halvorson, S., Sultana, F. and Laurie, N. (2009) "Introduction: global perspectives on gender-water geographies', *Gender Place and Culture* 16(4): 381–85.

Pain, R. (1991) "Space, sexual violence and social control," *Progress in Human Geography* 15(4): 415–31.

Penner, B. (2010) "Flush with inequality: sanitation in South Africa," The Design Observer Group. Available online at: http://places.designobserver.com/entryprint.html?entry=21619 (accessed May 16, 2013).

Runyan, A. S. and Marchand, M. H. (2000) "Conclusion: feminist approaches to global restructuring," in A. S. Runyan and M. H. Marchand (eds) *Gender and Global Restructuring: Sightings, Sites and Resistances*. London: Routledge, pp. 225–30.

Smith, L. (2004) "The murky waters of the second wave of neoliberalism: corporatization as a service delivery model in Cape Town," *Geoforum* 35: 375–93.

Sultana, F. and Loftus, A. (2011) *The Right to Water: Politics, Governance and Social Struggles*, London and New York: Taylor and Francis.

Swyngedouw, E. and Heynen, N. (2003) "Urban political ecology, justice and the politics of scale," *Antipode* 35(5): 898–918.

Undie, C., John-Langba, J. and Kimani, E. (2006) "'The place of cool waters': women and water in the slums of Nairobi, Kenya," *Wagadu: A Journal of Transnational Women and Gender Studies* Special Issue, 3. Available online at: http://appweb.cortland.edu/ojs/index.php/Wagadu/article/view/258/480 (accessed May 24, 2013)

UNICEF/WHO 2012 (2012) "Progress on Drinking Water and Sanitation," a Joint Report of the United Nations Children's Fund (UNICEF) and the World Health Organization (WHO). Available online at: www.unicef.org/media/files/JMPreport2012.pdf (accessed May13, 2013).

Valentine, G. (1992) "Images of danger: women's sources of information about the spatial distribution of male violence," *Area* 24(1): 22–9.

van der Hoek, W., Evans, B. E., Bjerre, J., Calopietro, M. and Konradsen, F. (2010) "Measuring progress in sanitation," in *Reaching the MDG Target for Sanitation in Africa*, Copenhagen: Danish Ministry of Foreign Affairs, pp. 42–50.

Wambua S. (2004) "Water privatization in Kenya," Global Issues Papers, No. 8, Heinrich Böll Foundation.

WSP (2004) "From hazard to convenience: towards better management of public toilets in the city of Nairobi, Water and Sanitation Program-Africa," Field Note. Available online at: www.wsp.org/wsp/sites/wsp.org/files/publications/329200792104_afFromHazardToConveniencePublicToiletsNairobi.pdf (accessed May 16, 2013).

WSP (2012) "Economics of sanitation initiative," A Report of the Water and Sanitation Program (WSP). Available online at: www.wsp.org/wsp/content/economic-impacts-sanitation (accessed May 16, 2013).

Yeboah, I. (2006) "Subaltern strategies and development practice: urban water privatization in Ghana," *The Geographical Journal* 172(1): 50–65.

2 "Out of the kitchen"

Gender, empowerment and microfinance programs in Sri Lanka

Seela Aladuwaka and Ann M. Oberhauser

Introduction

Microfinance programs claiming to reduce poverty through support for entrepreneurial activities are the focus of ongoing discussions in the field of development studies. Some analysts highlight positive impacts of these loan programs, emphasizing their ability to assist small-scale borrowers and provide financial support to low-income households in many areas of the developing world (Kabeer 2000; Fernando 2006). Others argue that these programs lead to increased poverty as borrowers become further indebted to lending institutions, thus impeding their livelihood strategies (Rankin 2001; Young 2010a). A common theme in most analyses of microfinance concerns the ability of these programs to address widespread gender inequities in developing regions through the empowerment of marginalized women (Fernando 2006; Pitt *et al.* 2006).

This chapter draws from feminist geography and development studies to analyze the institutional and socio-spatial aspects of microfinance programs in their efforts to support women's participation in economic activities, especially at the regional and community levels. In this analysis, the term "institution" is used in reference to the organizational structure, membership and local context of these programs, as well as their status as private, public or non-governmental organizations (NGOs). The discussion refers to empowerment as a means of breaking down existing gender stereotypes to facilitate women's increased participation in income-earning activities (Kabeer 2002). More specifically, women's empowerment enhances their ability to exercise power in the social institutions that govern their daily lives, such as the household, extended family, local community, markets and government (Carr *et al.* 1996). This approach to empowerment is useful in conceptualizing unequal social relations of gender and advocating for the redistribution of power among men and women (Pearson 1992).

Although community-based development and empowerment are often stated as key goals of microfinance institutions, their organization and impacts do not necessarily lead to these outcomes. This chapter argues that gendered social relations and spatial processes play important roles in these lending institutions. In particular, socio-spatial context, mobility and scale are geographic processes that inform our understanding of gendered economic strategies and institutional frameworks of microfinance programs. The study focuses on institutions that not only lend money to women, but also provide other services and support within specific social and spatial contexts. Three models of microcredit program that finance these economic activities, investments and basic subsistence services are examined here in order to compare and assess their efforts to effectively lend to and empower women.

The chapter is organized into five sections that address the conceptual framework, analyze the empirical aspects of microcredit programs in rural Sri Lanka and, finally, develop conclusions that inform broader analyses of gender and economic empowerment. The first section outlines a conceptualization of the role and impact of microfinance programs in developing regions through a feminist geography perspective. This approach emphasizes gender dynamics and socio-spatial aspects that impact the economic processes for both borrowing and lending money. The second section examines three microfinance programs in Sri Lanka that exemplify a variety of institutional structures and scales of participation in one state-owned and two non-governmental programs. The discussion focuses on the state-owned microfinance institution, Samudhiri Bank (SB), which has adopted a small-group approach to lending found in many microfinance programs where women are the majority of credit borrowers. The third section provides a background to the case study region of Kandy District, Sri Lanka, with a focus on how this microcredit institution works within communities to empower women and address issues of poverty and economic marginalization. Fieldwork conducted in this region yields substantial insight about gendered practices of credit programs among both borrowers and non-borrowers. The fourth section analyzes the socio-spatial dynamics and gender relations of microcredit programs in Sri Lanka with a focus on geographical context, scale and mobility among members of microcredit programs in this developing region. The conclusion offers critical insight to the socio-spatial impact of these lending institutions on women's mobility and the reconfiguration of social relations within domestic, community and regional spaces.

Microfinance: A feminist geography perspective

Many development scholars and practitioners have examined the role of microfinance in the context of socio-cultural norms and community-based economic strategies (Mahmud 2003; Rankin 2002; Weber 2004; Young 2010a). This chapter focuses on socio-spatial context, mobility and scale as important aspects of the operationalization and success of microfinance institutions for rural women. The discussion employs a feminist geography perspective in order to analyze how various scales or levels of involvement in microcredit activities influence, and are impacted by, both cultural and material aspects of individuals and households.

Scale and socio-spatial context are important aspects of household gender relations, since they impact access to, and distribution of, resources such as those required in microfinance activities (Kabeer 2000; Mahmud 2003; Pitt *et al.* 2006). Rankin's (2003) study of women's economic strategies in Nepal demonstrates the significance of scale through her analysis of the intersection of spatial practices, economic strategies and gender in the success and impact of microcredit programs. Specifically, she examines how local "economies of practice" in Nepal establish gendered ideologies that frame material opportunities differently for men and women. In another project, Ruwanpura (2007) analyzes the gender and ethnic constraints within which Sri Lankan microfinance NGOs operate, especially in cases where women are largely confined to certain domestic spaces and reproductive roles. Their involvement in income-generating activities supported by these NGOs gives them expanded use of space and increased access to resources. In addition, the role of communities, the state and international institutions in microfinance development vary widely depending on the area's socio-economic and political context.

Debates about the types of institutions involved in microfinance, such as NGOs, public or state entities, private institutions, community groups and intergovernmental organizations, reflect the contested and dynamic aspects of these lending programs (Roy 2010; Ringmar and Fernando 2006; Weber 2004). For example, Roy's (2010) analysis of global institutions that control the circuits of capital under which microfinance operates, provides insight to the lived experience of poverty in developing countries.

As argued here, the mobility of borrowers affects their impact on and participation in microfinance activities. Feminist geography highlights how increased mobility among women leads to greater resources and opportunities, while cultural attitudes and practices that dictate mobility can either hinder or expand their participation in microfinance. An abundance of feminist work on the intersection of mobility and microfinance examines the role of institutions and actors in the economic, social and political spaces of these development efforts (Mandel 2004; Worthen 2012; Wrigley-Asante 2012; Young 2010b). Socio-spatial context, mobility and scale are key geographic concepts employed in this chapter's analysis of gender and microfinance programs in Sri Lanka. This approach illustrates a shift in development studies from an approach that is static and economically biased to a more locally situated and socially derived analysis of institutions that shape microfinance in the development process.

Institutions of change: Microfinance programs in Sri Lanka

The expansion of microfinance programs in the developing world is an important part of poverty alleviation efforts and in many cases is designed to empower women. These programs are often implemented by governments, NGOs, community-based organizations, private banks and other institutions, which provide loans and credit to the poor in order to improve economic conditions as well as to raise the social status of borrowers (Roy 2010). However, local social and spatial histories are often overlooked in critically analyzing how these programs achieve their goals. In Sri Lanka, the development of microfinance as an approach to alleviate poverty has expanded due primarily to the growth of lending institutions and the mainstreaming of microfinance as a major development strategy and poverty alleviation program in the country (Tilakaratne *et al.* 2006). These efforts led to increased membership and participation of diverse groups in microfinance programs. By the 1990s, programs continued to expand and to recognize the provision of credit for poverty alleviation through international, national and local NGOs. Currently, a large number of institutions provide financial services to low-income groups in Sri Lanka (Tilakaratna 2005). Some of the problems faced by many low-income borrowers in particular include lack of basic inputs and support to establish small businesses, inadequate facilities, poor transportation and the inability to repay loans and to market their products.

This study is based on analyses of three microfinance programs in Sri Lanka, with specific attention to Samurdhi Bank, a state-owned poverty alleviation program (Table 2.1). The organizational structure, membership and size of these programs are discussed below to better understand their impact as microcredit development programs. Established in 1996, Samurdhi Bank encourages women to become active in small-group credit programs. More than 68 percent of credit borrowers in this program are low-income women. The second program, Sarvodaya Economic Enterprise Development Services (SEEDS), is an economic wing of Sarvodaya, the largest NGO in Sri Lanka, with women comprising over 60 percent of its members (SEEDS 1999). This program includes a comprehensive savings and credit program at the village level in support of Sarvodaya's long-term objective of promoting economic self-development among the poor. Finally, the Janashakthi Bank, started in 1989, is a women-only credit program as well as an economic NGO within the Women's Development Federation. The goal of this bank is to reach women who need assistance, to eradicate poverty and to reduce discrimination in accessing credit.

Membership in all three institutions has increased since their establishment of microfinance programs. Samurdhi Bank has the highest membership, with over 2.5 million in 2009, while SEEDS had 1.2 million and Janashakthi had nearly 40,000 members in 2010 (Table 2.1) (Samurdhi Authority of Sri Lanka 2012; SEEDS 2012). In addition, Samurdhi Bank has the highest number of borrowers, with nearly 700,000 in 2008. Members are those who have joined the organization, while borrowers are those who are members who also receive loans from the organization. As indicated in Table 2.1, the most common types of loans among all three programs are those that provide support for income-generating activities in agriculture and small businesses and for meeting basic housing and emergency needs. Social mobilization (or building confidence in individual capacities), as well as the ability to advance standards of living among members, are important benefits to participants in these programs (Tilakarantna 1993: 2).

An important aspect of these microfinance programs is the small group organization that promotes mutual support and helps to overcome isolation among its members. This feature is important to the operation of these microcredit programs and, specifically, the focus and size of their loans. The Samurdhi Bank is organized into small groups of women who meet weekly to discuss each other's problems, build solidarity and share financial information about their work. Within the Samurdhi credit program, these groups have a savings component and an

Table 2.1 Profiles of selected microfinance programs

Category	Samurdhi Bank (SB)	Sarvodaya Economic Enterprises Development Services (SEEDS)	Janashakthi Bank (JB)
Type	State-owned National operations	Non-governmental organization National operations	Non-governmental organization Regional and local operations
Focus of loans	Self-employment Agriculture Fisheries Emergency funds Housing Industry	Income-generation Consumption loans to improve quality of life Expand small businesses to create employment	Small business Agriculture Housing Welfare and emergency funds
Category of program	Credit Savings	Credit Savings Micro-insurance	Credit Savings Micro-insurance
Members	2,588,059 (2009)	1,223,442 (2009)	39,761 (2010)
Borrowers	698,417 (2008)	178,509 (2009)	15,791 (2010)
Branches/ banking societies	1,042 (2008)	65 (2008)	99 (2010)
Loan size	Maximum–100,000 rupees	Maximum–500,000 rupees	Average–6,084 rupees

Sources: SEEDS (2003, 2004, 2008); Women's Development Federation (2010); Samurdhi Authority of Sri Lanka (2012).

intra-group credit system, which promotes saving among poor households as a means of improving their financial management skills (Table 2.1). They also serve as an entry point to social mobilization and a focus for collective action. The majority of Samurdhi Bank's assets are made up of share capital in which all group members have to purchase 500 rupees in order for them to be a member of the bank.

Although the small group organization is meant to build solidarity among members, tension sometimes arises when individual members do not repay their loans on time and are required to pay additional interest rates. Another drawback to this model of microfinance is the small amount of money available to borrowers. According to data collected about the Samurdhi program, the loans are usually insufficient to start or to continue viable micro-enterprises (Aladuwaka 2003). Many of the credit borrowers in this program feel that if they have access to more money, they could invest in something more sustainable than a small vegetable garden, for example. Some of the respondents also said they had difficulty earning enough money to repay their loans, and, therefore, they had to find other work to supplement their earnings.

The second program, SEEDS, also operates through small groups that report to Sarvodaya Village Societies (Table 2.1). Similar to Samurdhi, once clients are members, they form small groups and are eligible for loans. Within this structure, projects target marginalized and impoverished women in Sri Lanka through the combination of small loans and training schemes. Finally, Janashakthi Bank is a member-based microfinance NGO designed to help the poor maintain their own economic activities by providing loans to women as a means of improving the socio-economic status of families (Charitonenko and de Silva 2002). According to the Women's Development Federation (1991), the strength of Janashakthi lies in the small group organization of its members. In addition, loans are given to the female members, whose involvement in all aspects of the program leads to empowerment through increasing their knowledge and self-confidence (United Nations Economic and Social Commission for Asia and the Pacific 1998). The umbrella organization to Janashakthi Bank, the Women's Development Fund, covers eight out of twelve administrative divisions in Hambantota District and one division in Monaragala District, reaching 740 villages in the southern coastal region of Sri Lanka (Women's Development Federation 1991) (Figure 2.1).

Although their organizational structures and types of membership differ, these three microfinance programs are designed to provide credit to marginalized people, thus assisting communities in many parts of Sri Lanka. The Samurdhi credit program focuses on credit and savings, while SEEDS is more comprehensive

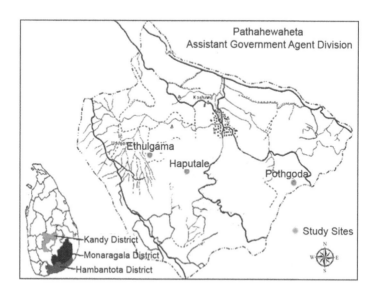

Figure 2.1 Study sites in Kandy District, Sri Lanka. (Fieldwork by Seela Aladuwaka, 2002.)

because it not only provides credit but also offers other non-financial services such as training, marketing and technical support to its clients (Charitonenko and de Silva 2002). Finally, by targeting women from poorer families, Janashakthi Bank promotes social mobilization and attempts to address isolation in economically deprived districts by establishing a small group organization, while improving the economic status of its members. These different models of microfinance affect the socio-spatial impact, and thus the empowerment of women, in diverse ways.

Microfinance programs in Sri Lanka: a comparison of borrowers and non-borrowers

The previous discussion of microcredit programs provides an important background to this analysis of how these lending institutions impact communities and households. This section focuses on the Sri Lankan case study region of Kandy District, where the Samurdhi credit program has operated for many years. Nearly one-third of the families in Kandy District are members of the Samurdhi Bank, making its examination relevant to understanding the effectiveness of microcredit programs in an impoverished and rural area. According to the 2010 census, Kandy is one of the most populous districts in Sri Lanka, with more than 1.4 million or 6.5 percent of the country's 21.5 million people. Over 85.9 percent of the people in this district live in rural areas (Central Bank of Sri Lanka 2011) and 64 percent of households are in debt. Thus, poverty is a major concern for many families in this area. Three localities, Ethulgama, Haputale and Pothgoda, in the Pathahevahata Assistant Goverment Agent (AGA) division are the specific study sites for this project (Figure 2.1).

In this research project, both primary and secondary data are used to analyze the socio-spatial aspects of microfinance programs. Primary data were gathered through field research on the Samurdhi Bank during nearly five months in 2002, while information about SEEDS and Janashakthi Bank provided important secondary data for this analysis. The fieldwork in Kandy District included semi-structured questionnaires, focus group discussions, participant observation and gender activity profiles (Aladuwaka 2003). Thirty female credit borrowers who received loans under the SB program and thirty females who did not borrow money from credit programs, were selected for this study using a snowball-sampling method in three villages. The selection of both credit borrowers and non-borrowers was instrumental in comparing and evaluating the role and impact of micro-enterprise credit programs on women's empowerment.

Semi-structured questionnaires were administered to gather information about participants' involvement in, and experiences with, credit programs. Their responses included details about the amount of money they borrow from Samurdhi Bank, as well as the frequency and use of borrowing and information on repayments of these loans. The questionnaires also addressed the extent of these women's micro-enterprise work, including materials, marketing, problems and future plans. Many of the questions were open-ended and responses revealed a range of information on: changes in access to and control of resources; decision-making at the individual, household, community and organizational levels; self-esteem; and control of earnings.

Additionally, focus groups were used to gather information about borrowers' experiences with Samurdhi Bank. Discussion in these groups addressed women's experiences with the credit programs, family perceptions of their activities and advantages and disadvantages of participation in the programs. The fieldwork also entailed participating in (and observing) women's activities in several phases, which included talking to the Samurdhi Bank staff, manager and credit borrowers in weekly meetings and in Village Society meetings. Finally, gendered activity profiles provided detailed information about the gender division of labor and consequences for households and communities. This information shows why women and men have certain tasks and what it means in terms of power relations and the position of women within the household.

The socio-economic status of the participants is somewhat similar to the general population. The majority of the participants come from low-income families with traditional gender roles in the household and low participation in the formal labor force. Additionally, the mean age of participants is 39 years and 90 percent of them are married. Comparing education and types of employment among the borrowers and non-borrowers highlights some of the benefits of women's involvement in microfinance. Education level is often linked to socio-economic status and reflects the potential to succeed in income-generating activities (Roy 2010). In our sample, education level differs slightly between borrowers and non-borrowers (Table 2.2). Approximately one-fourth of the participants (23 percent of borrowers and 26 percent of non-borrowers) have education levels that range from grades one to five. Moreover, 15 percent of borrowers and 10 percent of non-borrowers have a grade 12 education (field survey 2002). No one in the sample has a university degree.

In addition, the type of employment among participants also differs between the borrowers and non-borrowers (Table 2.3). Credit borrowers have a high rate of self-employment (37 percent) compared to non-borrowers (10 percent) and a relatively high percentage of borrowers engage in casual work (13 percent) compared to non-borrowers (3 percent).

Table 2.2 Educational level of respondents

Education	Borrowers (%)	Non-borrowers (%)
No schooling	6	7
Grade 1–5	23	26
Grade 5–10	33	30
10th grade (ordinary level)	23	27
12th grade (advanced level)	15	10
Degree	0	0
Total	100	100

Source: field survey (2002).

Table 2.3 Employment among borrowers and non-borrowers

Type of Employment	Borrowers (%)	Non-borrowers (%)
Farmer	17	13
Self-employed	37	10
Laborer (casual work)	13	03
No formal job	33	73
Total	100	99

Source: field survey (2002).

Finally, microcredit borrowers in the study invest their loans in various economic activities. Farming is the most common activity (with 43 percent investing in this activity), while 23 percent use their loans to buy material for housing. Moreover, 17 percent invest their money in establishing small businesses in their villages, a popular investment among women borrowers. The case study region thus provides a diverse context where borrowers are shown to engage in a variety of economic activities compared to non-borrowers. And, although poverty remains one of the main problems in the region, it has great potential to benefit from microfinance and overall investments in women and households. Are there differences in the benefits that borrowers and non-borrowers derive from microcredit programs? The next section examines the socio-spatial aspects of microcredit programs, with a focus on mobility of borrowers and non-borrowers in the rural Sri Lankan study site of Kandy District.

Geographies of credit and women's empowerment

Mandel (2004), Worthen (2012), Young (2010a) and other feminist geographers highlight the important role of spatial processes, such as mobility, in women's economic activities. Field data, detailed below, confirm that women's participation in Samurdhi Bank credit programs has given them more options for personal mobility than non-participating women and improves their overall confidence and ability to work for their own and their family's benefit. This mobility thus proves to be an important factor in women's empowerment, especially given the cultural parameters regarding travel among women. For example, rural women are less mobile due mostly to traditional roles linking them to household and domestic responsibilities.

In recent decades, however, women's migration to urban areas for work in factory-based jobs, and to the Middle Eastern region for work in domestic labor, has increased. Although the status of women in Sri Lanka is somewhat better than women in South Asia as a whole, traditional beliefs and practices continue to place limitations on women that impact their work opportunities (Aladuwaka and Oberhauser 2011). The majority of rural women engage in low-skilled, low-income economic activities in the informal sector and in unpaid agricultural labor (Jayaweera *et al.* 2007). Thus, gender divisions of labor, combined with traditional

Table 2.4 Mobility and decision-making among borrowers and non-borrowers

Decisions about travel	Borrowers (%)	Non-borrowers (%)
Individual decision	27	17
Husband informed of travel decision, but do not ask permission	40	22
Permission from husband required	20	36
Travel with husband	10	22
Travel with family member	3	3
Total	100	100

Source: field survey (2002).

beliefs whereby women are restricted from traveling alone outside their villages, considerably limit women's mobility.

According to this study, 27 percent of the borrowers stated that they could decide if they wanted to travel alone, especially to visit family or friends or to go outside the village (Table 2.4). In contrast, only 17 percent of non-borrowers made that decision without permission from their husbands. The number of women needing to get permission from their husband to travel is greater among non-borrowers than borrowers—36 percent compared to 20 percent respectively. Borrowers of credit, or participants in the microfinance program, also have more flexibility in deciding whether or not to travel with their husbands. As women become involved in the credit program, their ability to travel increases. Additionally, their participation in meetings outside the village, often translates to greater self-esteem. For example, a woman from Pothgoda who makes handbags and is very successful in her business stated:

> I get to go to all these meetings and I go to town very often to sell my bags and my business is growing. I am busier than ever before and happy to be part of the program. My husband never even made his own tea before, but now he knows I have important work outside of the home such as weekly meetings in Samurdhi Bank. He even cooks now which he did not before. He respects my work and involvement in this work.

Another borrower who runs a village boutique in Haputale remarked:

> I am happy that I now help family to get more income, and I am working in store. Because my work now I get chances to go out to the meetings outside the village. I feel I play an important role in family.

Thus, expanded opportunities that stem from women's decisions to travel to the Samurdhi Bank for transactions improve their communications skills with bank officials and overall autonomy in their lives.

Cultural norms, which limit the mobility of rural women, are challenged by their involvement in these entrepreneurial activities. In general, women's mobility increases as they attend regular meetings, while greater visibility in the community boosts their confidence. Similar to Rankin's (2001) work on microfinance in Nepal, this research suggests that through microcredit programs, women's spatial mobility has expanded their socio-economic practices and overall empowerment. Studies from Bangladesh also show that microcredit programs have increased mobility and strengthened networks among women who were previously confined to the home (Carr *et al.* 1996; Kabeer 2000). This aspect is critical to women in rural areas in Kandy District, where their mobility is limited.

The experiences of Samurdhi Bank credit borrowers reinforce the importance of intersecting scales of microcredit activities, from local to regional and national levels of both economic networks and mobility. In Kandy District, women borrowers travel outside their homes and villages to participate in monthly meetings or to obtain loans at the district bank (Figure 2.2). Credit borrowers also travel to nearby cities to purchase materials or to sell their products in town. One credit borrower (from Ethulgama) in our study takes her products to Kandy on a weekly basis, developing new ideas and expanding her business opportunities. Some women credit borrowers also have opportunities to travel from their villages to national meetings in the capital city of Colombo. Increased experiences outside their villages often improve these women's ability to network. However, they also face certain challenges as a result of this increased travel. When women are gone from home, more work is generated when they return. On some occasions, husbands help them in their domestic work but in most cases, women continue to maintain responsibility for domestic work.

In the Samurdhi credit program, women are more likely to exercise some autonomy in the social institutions that govern their lives. This empowerment reflects a significant difference between the borrowers and non-borrowers in terms of participation in community activities. Sixty-three percent of borrowers, compared to

Figure 2.2 Women receiving loans at the Samurdhi Bank. (Photo by Seela Aladuwaka, 2002.)

only 30 percent of non-borrowers, stated that they decide if they want to participate in community work. These activities give women the opportunity to travel outside their homes and to establish a strong social network. Increasing women's ability to leave their, sometimes isolated, situations to meet other women and build networks is an important aspect of empowerment. As one respondent stated, "Because we also have a group of five women, we have much help, we help each other and have some friends that we can go to if we need someone; that is strength, I think." This comment highlights that through their experiences with group participation in community activities, women develop a support system that they did not have before.

Specifically, we observed that women are able to talk about their personal problems, such as a husband's abuse of alcohol and domestic violence. Leaders in the groups have been able to advocate for these issues in ways that have impacted the villages. One of the Samurdhi society leaders in Pothgoda explained their efforts to combat men's alcohol problems in the village through public meetings. As indicated in the following quote, many women in this area believe these efforts have helped them to stand up to their husbands' alcohol related problems:

> Another thing we encourage women to do is stop their husband's alcohol problem. We try to raise awareness of this and try to reduce it. In Samurdhi meetings, we advocate the action against alcoholism and this helps us to get stronger in taking actions. If women come together, we can do things we could not do before.

As stated here, women's active roles in the community often leads them to demand that their husbands reduce their alcohol consumption:

> My husband was alcoholic and before he did not let me go to Samurdhi meetings. I told him he must stop drinking alcohol because as a group leader I cannot ask other women to take action against alcoholism if he does not change. I am happy that he now seems to understand it and I see some improvements, he does not drink as much as he used to. He does not make any troubles when I go to meetings and he even helps me out with my work. So, I can tell other women to do the same as I did.

This woman's ability to curb her husband's alcohol habit improved her situation and may encourage other women in the village to do the same.

In addition, participation and decision-making roles in Samurdhi Societies provide opportunities for women to become leaders in their communites. This research finds that women's involvement in leadership positions is significantly higher among borrowers (37 percent) in comparison to non-borrowers (6 percent). Women also hold various types of leadership positions in village societies, such as small group leaders, secretaries and treasurers. Most of the village Samurdhi Development Officers are women who make decisions regarding credit provisions to their villages. One thirty-year-old woman who has been an active leader in the credit program stated:

> Now men also expect us to work with them equally in banks, as well as other work. Our daughters learn from us in how to change the way we work, we can participate in development work rather than doing all the duties at home.

Another thirty-year-old respondent from Pothgoda stated:

> Because we can participate in the programs, it helps to change the belief that "women belong in the kitchen." This credit program opened us up to opportunities to come out from home. Otherwise, we are trapped inside the house, doing all the work. We never get to go out.

These quotes indicate that women identify ways in which these microfinance programs lead to empowerment by challenging their traditional roles and, in some cases, becoming role models for their daughters. These experiences are similar to those described by Mahmud's (2003) work in Bangladesh, where participation in microcredit is linked to expanding women's choices and decision-making at both household and community scales.

However, women's involvement in income-generating activities can present challenges when they step out of their traditional roles. Sinha claims that:

> traditional gender role expectations and patriarchal attitudes in many developing nations make it even more difficult for women to relieve themselves of family responsibilities. The familial and social conditioning in many developing countries inhibits the confidence, independence and mobility of women.
> (Sinha 2005: 1)

Although participation in credit programs sometimes conflicts with women's roles as primary caregivers (Worthen 2012), they benefit in ways that affect many aspects of their lives. For example, Kabeer argues that "this new form of consciousness arises out of women's newly acquired access to intangible resources of analytical skills, social network, organizational strength, solidarity and sense of not being alone" (1994: 246). Women's unrestricted mobility has the potential to improve their access to community networks, information and resources, thus increasing their economic participation and self-esteem. In this study, a forty-five-year-old credit borrower from Pothgoda village never traveled on her own before she joined the credit program. She now travels often to nearby towns to sell her handbags and is more motivated to continue her income-generating work with this newly found independence. Another thirty-seven-year-old mother of three in Haputale village, who continues to work in the village store stated:

> I am happy that I now help my family to get more income, and I am working in the store. I have never had an income earning before, and I feel I play an

important role in my family by helping them and I feel valued. Now, I get to meet people in the village as I work in the store all day long and I feel more confident. I get chances to go out from the home for all these meetings and even to the town.

As indicated here, women's expanded mobility has given them greater self-confidence through increased respect at home and in the community. Furthermore, women's involvement in credit programs helps them to move beyond their household labor and actively engage in society through their daily activities. In Sri Lanka, limited mobility among rural women (due to their domestic responsibilities) confines them to their villages and immediate surroundings. Thus, freedom of movement for women means expanding their spatial experiences and improving their knowledge base and networking. Many of these opportunities stem directly from their participation in microfinance.

With regard to other microcredit institutions in Sri Lanka, the organization and small group structure improves women's confidence and self-esteem (Hewavitarana 1994). According to Colombage's work on Janashakthi Bank, "small groups operating at the bottom of the organizational structure of the WDF have greatly enhanced women's empowerment and capacity building" (2004: 22). He also reports that small group meetings have given female members the opportunity to exchange knowledge and experience with regard to savings, credit and enterprise development. Similar to Samurdhi, Janashakthi Bank's organizational structure is based on a bottom-up approach, and its leadership, management and staff are drawn from the grass-root level of membership. Credit and savings are carried out by women who have mobilized and organized themselves and who are willing to take initiatives to improve their socio-economic conditions. In this process, "the poor women of Hambantota have become dedicated change agents in the alleviation of poverty and in developing their areas, generating jobs and creating confidence among themselves" (Tilakarantna 1993: 2).

Thus Janashakthi's organizational structure has given women opportunities to be leaders and decision-makers for themselves, their families and their communities. The bottom-up structure of Janashakthi Bank has broken the isolation of poor women, created a forum for interaction and given them social recognition (Colombage 2004). Thus, women members of this bank have an effective instrument to deal with outside agencies and bargain for resources and service deliveries. The impact of the bottom-up organizational structure has a positive outcome for women borrowers as "they are no longer atomised individuals; they have emerged as a social power of some sort to reckon with" (Tilakarantna 1993: 2).

In addition, the SEEDS microfinance program continues to expand its services across Sri Lanka, including war-torn areas of the north and east. As the majority of borrowers, women benefit from the comprehensive services provided by SEEDS at the grass-roots level. With its Enterprise and Training Divisions, borrowers get more support to carry out their enterprises effectively. In their study on SEEDS, Simon and Sear (2005) point out that loan recipients, especially women, claim that they are empowered as a result of their new responsibilities. Borrowers tend to benefit from their new social standing or status within the home and the local community.

The microfinance programs outlined in this analysis demonstrate how institutional frameworks intersect with socio-spatial processes to not only create opportunities, but also raise challenges for women to overcome in their economic strategies. The effects of these programs on both borrowers and non-borrowers in the state-owned microcredit lending program, Samurdhi Bank, are compared by measuring impacts such as mobility, access to income-generating activity, decision-making in their households and communities and developing leadership skills. In most cases, microcredit borrowers are much better off than non-borrowers in these aspects of empowerment and access to resources. However, positive outcomes among borrowers coincide with significant challenges, as women have fewer material resources, are less mobile and have trouble with competing demands on their time given their domestic roles and increasing economic activities (Rankin 2002; Young 2010b). These challenges to women who participate in microfinance programs often lead to contested arenas of power and resistance.

Conclusion

The socio-spatial dimensions of efforts to expand economic opportunities for women in developing regions require a critical analysis of the processes and places that underlie these opportunities. Feminist analyses recognize and support women's roles in the development process and their ability to generate income through community-based livelihoods. Microfinance programs often target marginalized women as a means of enhancing their economic status and thus benefitting their households and communities. This chapter draws from feminist geography and development studies in order to examine spatial processes that help define both the institutional context of microcredit programs and the forms of empowerment that result from women's participation in these programs.

Three microfinance programs in Sri Lanka, the state-owned Samurdhi Bank and two NGOs (SEEDS and Janashakthi Bank), form the basis for the discussion. The small group and village-based structure of the Samurdhi program tends to improve and support the status and self-confidence of borrowers. These experiences in turn promote women's empowerment through expanded socio-spatial aspects of their economic activities. Findings from this research on Samurdhi Bank borrowers indicate that involvement in credit programs has the potential to empower women by building capacity and increasing networks through greater mobility. The comparison of borrowers and non-borrowers provides information about the similarities and differences in women's social roles, education and employment options. Because their space is often confined to the domestic sphere as a result of cultural attitudes and roles, many women lack opportunities to network with other entrepreneurs outside their households and communities. As emphasized by Young, although mobility is influential in establishing contacts and expanding access to resources, it is "often determined by educational level or other indicators of socio-economic status" (2010a: 620). In addition, Young (2010a) argues that many women find that mobility is not necessarily positive, and they prefer the convenience of working from home. Therefore, given the opportunity to participate in microfinance programs, women engage in work and connect with the broader society in ways that fit their personal and familial needs.

According to many feminist and development scholars, microfinance institutions have the potential to play a greater role in transforming gender relations through increased mobility and disrupting unequal social relations (Shakya and Rankin 2008; Mahmud 2003; Fernando 2006). Mobility may not be the primary goal; however, the institutional structure of these programs can increase borrowers' access and connection to areas and people outside their immediate communities. In this case study, Samurdhi borrowers in Kandy are able to travel outside their villages, participating regularly in meetings and establishing networks. Our analysis finds that increased spatial mobility empowers women and improves their self-esteem because of their increased involvement in the public sphere.

Microfinance and micro-enterprise activities help women take control of their own lives, thus challenging traditional gender hierarchies that confine women to the home and reproductive labor. Experiences of microfinance borrowers help them not only to gain immediate needs, such as employment and income, but also to question gender inequalities that they experience in their daily lives (Mandel 2004; Raju 2005). Findings from this research contribute to our understanding of microfinance operations and policy development that lead to expanded opportunities for empowerment through participation in microcredit programs. Moreover, gender biases embedded in society are shown to limit women's mobility, social interactions, economic participation and access to business development services (Sinha 2005). Policy implications of this research highlight how microfinance institutions need to recognize and critically examine these fundamental issues when developing their programs. Thus, studies such as this provide contextual and empirical insights to larger issues embedded in development work. This analysis also emphasizes the need for more research about how microfinance intervention influences borrowers' socio-spatial experiences, and, in turn, the interrelationship between women's empowerment and development.

References

Aladuwaka, S. (2003) "Credit program, poverty alleviation and women's empowerment: a case study from rural Sri Lanka," dissertation, West Virginia University. Available online at: http://hdl.handle.net/10450/3129 (accessed May 20, 2013).

Aladuwaka, S. and Oberhauser A. M. (2011) "Contextualizing credit programs, poverty alleviation and women's empowerment: a case study from rural Sri Lanka," in Raju Saraswati (ed.) *Gendered Geographies: Interrogating Place and Space in South Asia*, India: Oxford University Press, pp. 245–67.

Carr, M., Chen, M. and Jhabvala, R. (1996) *Speaking Out: Women's Economic Empowerment in South Asia*, Canada: Aga Khan Foundation.

Central Bank of Sri Lanka (2011) "Economic and Social Statistics." Available online at: www.cbsl.gov.lk/pics_n_docs/10_pub/_docs/statistics/other/econ_&_ss_2011.pdf (accessed May 20, 2013).

Charitonenko, S. and de Silva, D. (2002) *Commercialization of Microfinance: Sri Lanka*, Manila: Asian Development Bank. Available online at: www.adb.org/publications/commercialization-microfinance-sri-lanka (accessed May 17, 2013).

Colombage, S. S. (2004) "Microfinance as an instrument for small enterprise development: opportunities and constraints," Occasional paper No. 52, Twenty-third Anniversary Lecture, Centre for Banking Studies, Central Bank of Sri Lanka.

Fernando, J. L. (1997) "Nongovernmental organizations, micro-credit, and empowerment of women," *Annals of the American Academy of Political and Social Science*, 554: 150–77.

Fernando, J. L. (ed) (2006) *Microfinance: Perils and Prospects*, New York: Routledge.

Hewaviharana, B. (1994) *Build Up a Bank and Grew With It. The Janashakthi Banku Sangam*, Hambantota: Women's Development Federation.

Jayaweera, S., Wijemanne, H., Wanasundera, L. and Vitarana, K. M. (2007) *Gender Dimensions of the Millennium Development Goals in Sri Lanka*, Colombo, Sri Lanka: Centre for Women's Research. Avaialble online at: www.lk.undp.org/content/srilanka/en/home/library/mdg/gender-dimensions-of-the-millennium-development-goals-in-sri-lan/ (accessed May 18, 2013).

Kabeer, N. (1994) *Reversed Realities: Gender Hierarchies in Development Thought*, London: Verso.

Kabeer, N. (2000) "Conflicts over credit: re-evaluating the empowerment potential of loans to women in rural Bangladesh," *World Development* 29(1): 63–84.

Kabeer, N. (2002) "Resources, agency, achievements: reflections on the measurement of women's empowerment," *Development and Change* 30: 435–64.

Mandel, J. (2004) "Mobility matters: women's livelihood strategies in Porto Novo, Benin," *Gender, Place, and Culture* 11(2): 257–87.

Mahmud, S. (2003) "Actually how empowering is microcredit?," *Development and Change* 34(4): 577–605.

Pearson, R. (1992). "Gender Matters in Development," in T. Allen and A. Thomas (eds) *Poverty and Development in the 1990s*, Oxford: Oxford University Press, pp. 291–312.

Pitt, M. M., Khandker, S. R. and Cartwright, J. (2006) "Empowering women with microfinance: evidence from Bangladesh," *Economic Development and Cultural Change* 54(4): 791–831.

Pitt, M. M., Khandker, S. R., Chowdhury, O.H. and Millimet, K. L. (2003) "Credit programs for the poor and the health status of children in rural Bangladesh," *International Economic Review* 44(1): 87–118.

Raju, S. (2005) "Gender and empowerment: creating 'thus far and no further' supportive structures, a case from India," in L. Nelson and J. Seager (eds), *A Companion to Feminist Geography*, Malden, MA: Blackwell, pp. 194–207.

Rankin, K. (2001) "Governing development: neoliberalism, microcredit, and rational economic woman," *Economics and Society* 30/1: 18–37.

Rankin, K. (2002) "Social capital, microfinance, and the politics of development," *Feminist Economics* 8(1): 1–24.

Rankin, K. (2003) "Cultures of economics: gender and socio-spatial changes in Nepal," *Gender, Place and Culture* 10(2): 111–29.

Rankin, K. (2006) "Social capital, microfinance, and the politics of development," in Fernando J. (ed.) *Microfinance: Perils and Prospects*, New York: Routledge, pp. 77–96.

Ringmar, E. and Fernando J. (eds) (2006) *Microfinance: Perils and Prospects*, New York: Routledge.

Roy, A. (2010) *Poverty Capital: Microfinance and the Making of Development*, New York: Routledge.

Ruwanpura, K. N. (2007) "Awareness and action: the ethno-gender dynamics of Sri Lankan NGOs," *Gender, Place and Culture* 14(3): 317–33.

Samurdhi Authority of Sri Lanka (2012) "Samurdhi Bank Societies." Available online at: www.samurdhi.gov.lk/web/ (accessed May 17, 2013).

Sarvodaya Economic Enterprise Development Services (SEEDS) (1999) "Annual Report 1998–1999," SEEDS, Sri Lanka.

Sarvodaya Economic Enterprise Development Services (SEEDS) (2003) "Annual Reports 2003–2005," SEEDS, Sri Lanka.

Sarvodaya Economic Enterprise Development Services (SEEDS) (2004) "Annual Report 2004," SEEDS, Sri Lanka.

Sarvodaya Economic Enterprise Development Services (SEEDS) (2008) "Sarvodaya Economic Enterprises Development Services." Available online at: www.seeds.lk/ (accessed May 16, 2013).

Sarvodaya Economic Enterprise Development Services SEEDS (2012) "BWTP Member Profile: Sarvoday Economic Development Services. Available online at: www.seeds.lk/ (accessed 7 July 2012).

Shakya, Y. B. and Rankin, K. N. (2008) "The politics of subversion in development practice: an exploration of microfinance in Nepal and Vietnam," *Journal of Development Studies* 8(44): 1214–35.

Simon, D. and Sear, J. (2005) *Indigenous Microcredit and Enterprise Establishment: A Sri Lanka Case Study*, Hyderabad, India: ICFAI Press.

Sinha, S. (2005) *Developing Women Entrepreneurs in South Asia: Issues, Initiatives and Experiences*, Bangkok: UNESCAP. Available online at: www.unescap.org/tid/publication/indpub2401.pdf (accessed May 17, 2013).

Tilakaratna, G. M. (2005) *Microfinance in Sri Lanka: A Household Level Analysis of Outreach and Impact on Poverty*, Colombo, Sri Lanka: Institute of Policy Studies.

Tilakaratna, S. (1993) *Social Banking to Meet the Needs of the Poor*, Hambantota, Sri Lanka: Women's Development Federation.

Tilakaratne, G., Galappattige, A. and Perera, R. (2006) "Promoting empowerment through microfinance in Sri Lanka," *Economic and Political Empowerment of the Poor (EPEP): Country Studies of Sri Lanka*, SACEPS Paper No. 9. Kathmandu: South Asia Centre for Policy Studies.

United Nations Economic and Social Commission for Asia and the Pacific (1998) "A bottom-up approach towards poverty alleviation." Available online at: www.escap-hrd.org/csri.htm (accessed May 17, 2013).

Weber, H. (2002) "The imposition of a global development architecture: the example of microcredit," *Review of International Studies* 28(3): 537–55.

Weber, H. (2004) "The new economy and social risk: Banking on the poor," *Review of International Political Economy* 11(4): 356–86.

Women's Development Federation (1991) *A Report on Janashakthi*, Hambantota: Women's Development Federation.

Women's Development Federation (2010) Available online at: www.jwdf.org/ (accessed May 17, 2013).

Worthen, H. (2012) "Women and microcredit: alternative readings of subjectivity, agency, and gender change in rural Mexico," *Gender, Place and Culture* 19(3): 364–81.

Wrigley-Asante, C. (2012) "Out of the dark but not out of the cage: women's empowerment and gender relations in the Dangme West district of Ghana," *Gender, Place and Culture* 19(3): 344–63.

Young, S. (2010a) "Gender, mobility and the financialization of development," *Geopolitics* 15(3): 606–27.

Young, S. (2010b) "The moral hazards of microfinance: restructuring rural credit in India," *Antipode* 42/1: 201–23.

3 Neoliberalization, gender and the rise of the diaspora option in Jamaica

Beverley Mullings[1]

Introduction

In the last fifteen years, interest in the developmental impact of transnational exchanges within migrant networks has grown among international development institutions and practitioners. This growing attention to the developmental contributions of diaspora represents a significant shift in the value and importance that states and international development institutions have begun to accord to migration processes and, increasingly, migrants themselves. While development scholars and policy makers in the 1970s and 1980s saw the relationship between migration and social transformation in largely negative terms, these views were reversed by the mid 1990s; the increase in recognition of the magnitude and reliability of the economic flows generated by migrants led to a resurgence of interest in the developmental role that migrant communities have the potential to play.

The diasporic turn in development policy circles represents a renewed optimism in the relationship between migration and development—an optimism that is based on the importance of the contribution of émigrés to the economic successes of countries like Israel, India and China (Kuznetsov 2006; Ionescu 2006). Jamaica was one of the early adopters of the "diaspora option," a term that policy makers at the World Bank have used to describe a broad set of policy initiatives aimed at utilizing the economic, human and social capital of populations abroad to revitalize growth and development (World Bank 2007). As early as 1993, Jamaica began to develop the policy infrastructure needed to facilitate a greater level of participation among its diaspora. These policies evolved to include the creation of a government department dedicated to diaspora affairs, a Diaspora Advisory Council, a Diaspora Foundation and research institute and a biennial conference where Jamaicans abroad propose, consider and strategize future developmental interventions (see Mullings 2012; Sives 2012). As a result of these early efforts, Jamaica is often seen as a model among countries seeking to develop their own diaspora strategy.

As diaspora options evolve, states are increasingly shifting their development policies away from maintaining inflows of migrant remittances and toward capturing the investments and knowledge networks of members of their diaspora. The shift in government discourse from "remitting migrants" to "an investing diaspora" is both discursive and material; along with the changing terminology, state interests have moved away from the incorporation of remitting migrant populations (concentrated among the lower middle and working classes) toward professional, entrepreneurial and highly skilled groups who are viewed as more likely to generate economic growth at the community and national levels (Kuznetsov 2006). But, as I shall argue in this paper, the shift from migrant to diaspora obscures the way that gender structures diasporic formations and influences the durability of diasporic flows. Without critical reflection on the relationship between gender, migration, diaspora and development, states in both home and host countries are likely to reproduce already existing gendered, racialized and classed relations within diaspora. Drawing on a case study of Jamaica and the deliberations that have emerged from a specific component of its diaspora option (the biennial conference), I examine how the gendered contributions of women within migration circuits are considered by states as they seek to incorporate populations abroad into their development strategies.

Shuval argues that "Diaspora is a social construct founded on feeling, consciousness, memory, mythology, history, meaningful narratives, group identity, longings, dreams, allegorical and virtual elements all of which play a role in establishing a diaspora reality" (2000: 43). Safran (1991) and Sheffer (2006) also use the term diaspora to describe ethnic minorities of migrant origin who live in host countries, but who maintain active emotional and material links with their countries of origin or ancestry. Brubaker (2005) observes that definitions of diaspora are dynamic and have shifted away from assimilationist and immigrationist paradigms that define diaspora as a bounded entity, to definitions that emphasize the de-territorialized stances and claims of populations who share an identity and distinct categories of practice. Like Brubaker, I argue that paying attention to specific claims and stances of such groups offers the greatest opportunity for assessing their contributions of social transformation, but I complicate his implicit assumption that diasporic stances are always visible or intelligible. For example, the diasporic identities, loyalties and mobilized energies of migrant populations that continue to motivate migrant remittances do not necessarily take forms that directly contribute to a diaspora project, yet they have proved crucial to maintaining and recreating cultural, ethnic and economic connections across territorial borders.

In this paper, I therefore expand our understanding of diaspora to include the emotional spaces created by the regular and systematic enactment and exchange of knowledge, commodities, economic resources and cultural practices among members of a nation dispersed across international territorial boundaries. These spaces are crucial to the creation and durability of diaspora because they are spatial manifestations of the emotions that migration, separation and alienation generate. In the context of an increasingly feminized and precarious global division of labor, the space of diaspora is increasingly a space within which women play a significant contributory role.

Gender, migration, diaspora and development

Only in the last twenty years has the gendered nature of relationships between migrants and their home countries become a significant topic of study among development scholars. This is not to say that development scholars have never considered the relationship between migration and social transformation, but rather that, historically, studies were based on macro-level accounts that constructed the migrant as a largely genderless, rational economic figure responding predictably to economic opportunities absent in the home country. How scholars made sense of the developmental outcomes of migration also depended on the theoretical frameworks that they supported. While modernization scholars like Lewis (1954), drawing on neoclassical frameworks, viewed migration as a positive process that brought material progress, by the 1970s and 1980s, others took a much more pessimistic position. Influenced by Marxist political economy, scholars like Paul Singer believed that the structural constraints faced by peripheral regions fuelled forms of migration that ultimately had a negative impact on both rural and urban communities. Like neoclassical approaches, most Marxist-inspired accounts paid more attention to the social transformations occurring in the places affected by migratory flows rather than those taking place among the people generating the flows themselves.

Writing in the late 1990s, Silvey and Lawson (1999) argued that the tendency to not centrally place the migrant within developmental frameworks contributed to a degree of narrowness in the ways that development scholars conceptualized and studied mobility across borders. This narrowness ultimately occluded the relations of power bound up in the migration process, their effect on migrant identities and patterns of mobility and the developmental outcomes generated both at home and abroad. Almost a decade later, Campt and Thomas (2008) made a similar observation in relation to the growing popularity of diaspora as a site of scholarship. They similarly warned scholars of the need to recognize the:

> asymmetrical relations of power that structure the dynamics of diaspora not only externally through the pressures that produce movement and migration, but also internally through the ways they configure complex relations of settlement, and racial and gendered formations within diasporic communities.
>
> (Campt and Thomas 2008: 3)

Since the 1980s, feminist contributions have highlighted the importance of gender to contemporary migration circuits. Drawing on political economy frameworks, scholars have drawn connections between the growing incorporation of women into global production circuits (Elson and Pearson 1981; Safa 1995; Wright 2006) and the increasing number of women entering migration circuits that were once largely associated with men (Momsen 1992; Pessar and Mahler 2003; Rodriguez 2010). As noted in a recent report of the UN Commission on Population and Development (United Nations 2006), in 2006, women accounted for approximately 49.6 percent of the world's migratory flows—with some countries recording proportions as high as 70 to 80 percent. Studies in the mid to late 1990s also showed that migrants did not lose contact with their home countries after migration but instead, maintained close ongoing relationships with communities, friends and family in their home countries (Portes 1995; Smith 2000; Thomas-Hope 1999). Through the maintenance of remittance-led transnational networks, migrants played an instrumental role in the economic and social welfare of families and communities, particularly during the early years of neoliberalization, when many governments in the global South retreated from the provision of welfare (Burman 2002; Glick Schiller and Fouron 2001; Itzigsohn *et al.* 1999).

The feminization trends in migration respond to a gendered international division of labor that relies upon women in the global South filling gaps in the social reproduction of economies and societies in the global North (Ehrenreich and Hochschild 2002; Kofman 2012; Erel 2012). Erel (2012) identifies the increasing labor market participation of women in the global North, longer life expectancy levels and, consequently, higher care needs as the main factors behind the care deficit in Europe. But she also sees the growing commodification of care through financial instruments, such as tax credits, as a contributory factor to these feminized transnational circuits. The benefits and costs of the feminization of migration are both contradictory and complex. While women's migration and incorporation into labor markets in the global North has been instrumental to the alleviation of poverty in their households and communities, women's ability to financially support their families and communities has often come at the expense of their ability to be a present source of emotional support and protection—particularly for their most vulnerable members.

The growing reliance in the global North on women from the global South to provide these services cannot be understood without reference to the role of states (Maher 2004; Neumayer 2006; Rodriguez 2010). For in both receiving and sending countries, states continue to play a significant role in shaping the value attributed to these forms of labor. In high labor-sending regions like Jamaica, women's migration and participation in caring labor markets has been invaluable to the inflow of foreign capital, while in high labor-receiving states like the United States, migrant women's participation in the care sector, often within informal and precarious settings, continues to reduce state spending within these domains. Yet, despite the importance of care work to the coffers of sending and receiving states, few governments acknowledge the exploitative nature of some of these forms of work, and even fewer actively seek to protect immigrant women's labor. Feminist scholars argue that it is precisely the socially reproductive nature of these forms of work that lie behind low levels of pay, high levels of control and weak labor protection (Katz 2001; McDowell 2009). But, as I argue in this paper, it is also the failure of states to value the socially reproductive importance of emotional and caring labor that continues to devalue the contributions made by women migrants (many of whom are concentrated in precarious work) to the durability of diaspora and sustained development. By unpacking the relationship between emotion, social reproduction and diasporic formulation, I fill the gap in emerging literature on diaspora that does not address the importance of gender to the durability of diasporic formation or its potential to bring about progressive change.

Women, the transnationalization of care and creation of diaspora in Jamaica

Over half a century of the incorporation of women into global care circuits—primarily as nurses and domestic workers—has significantly transformed gender relations among women and between women and men in Jamaica. In the 1950s, Jamaican women were drawn into the UK labor market in order to largely provide support or auxiliary professional services in the post-war economy (e.g. nursing, clerical and secretarial work). While large numbers of women migrated, only a few took their children with them. In fact, figures presented by Lowenthal suggest that the mass migrations of the 1950s may have also marked the beginning of the transnational circuits of care that have become a characteristic of Caribbean migration and diaspora formation (1972: 220). While migrating to Britain between 1955 and the 1960s, adults took 6,500 children with them, leaving approximately 90,000 behind.

Later migration circuits to North America drew large numbers of women into domestic service, and, in doing so, deepened the transnationalization of the Jamaican family and the familial bonds that lie at the heart of diaspora. This is largely because women who worked in domestic service were usually required to leave their own children in the care of extended family members and friends in order to look after the children and elderly relatives of their employers (Fog Olwig 1999; Bauer and Thompson 2006). Pratt (2009) describes women's migration, the trauma of family separation and the maternal loss that many women encounter when leaving their children behind as a form of circulating sadness due to transnational circuits of absence, yearning, responsibility and care. Burman (2002) also alludes to the sense of responsibility, attachment and guilt that women feel, as evidenced by the money and goods that they send to sites and people left behind as a form of reparation. But, as Trotz (2006) observes, these circulating yearnings have also been instrumental to the strength and durability of diaspora formations, because of the material benefits that these transnational flows of capital, care and emotion also generate. Therefore, I concur with Karen Fog Olwig (2007) that Caribbean family ties, forged through interpersonal relationships and within intimate spheres, lie at the heart of the processes that have given rise to diasporic formations. The high incidence of single parenting has been a significant factor in women's participation in international migration circuits and the intensity of the material and emotional flows across international borders that have ensued. It is the collective effort to overcome the drag of distance, the trauma of separation and the potential for forgetting that animates the flows of capital, material goods and ideas that constitute the "magic" of diaspora upon which so many governments and development institutions seek to capitalize.

The shift within development circles, from a general level of apathy toward questions of migration during the 1980s to the current enthusiasm toward diaspora among states since the 1990s, reflects the importance of migrant emotional geographies to state governing imaginaries. The shift in focus among states and international development institutions from migrant to diaspora is indicative of a growing awareness of the power of emotion in the production of transnational flows, but there has also been scant attention to the ethical questions that emerge when emotion becomes a governing technology of the state.[2] Therefore, it is important to make visible the material relationships that give rise to the emotional flows that create and sustain diaspora and their imbrication in broader economic and political systems of insecurity and violence. It is also important to talk about the role of states and international development institutions in shaping the economic, social and political landscapes within which the transnational emotional geographies so integral to diasporic formations are produced.

Women, economic crisis and the transnationalization of social reproduction in Jamaica

The number of Jamaicans living abroad is presently estimated to be approximately five million, and this reflects the island's long history of migration. Concentrated in the United Kingdom, the United States and Canada, the growth of the Jamaican diaspora reflects the island's persistent history of poverty and the continuous search among Jamaicans for "a better life." Jamaican women actively participated in international migration streams as early as the 1950s, when many moved to the United Kingdom to fill job vacancies that no longer attracted workers among the native British. Subsequent waves of migration to the United States and Canada saw women become a significant migrating population. While women in the 1950s constituted, on average, 40 percent of the migrating population, their numbers grew significantly in subsequent years, and by the 1970s, they accounted for 52 percent and 67 percent of immigrants to the United States and Canada respectively (IOM 2007). Although they have constituted more than half of all Jamaican migrants to the United States and Canada since the 1970s, women have done so at significantly different rates. Women's participation in US migration streams increased each decade. In Canada, however, it declined each decade after the 1960s. The declining proportion of women in Canadian migration streams over the last fifty years highlights the important role that states play in shaping the intensity and durability of the transnational circuits of emotion, care and material resources that migration sets in motion. In the case of Canada, while making provisions for Jamaican women to fill the care deficit created by the entry of Canadian women into the labor force, the Caribbean Domestic Scheme did so under immigration rules that unfairly barred those women and their children from entitlement to full immigration status (Trotz 2007). During 1977, campaigns to fight the deportation of seven Jamaican women, who had attempted to sponsor their previously undisclosed children on their applications for permanent residency, highlighted the implicit instrumentality of the state's demand for disembodied workers, who were expected to dutifully care for the families of Canadians without regard for their own (Fudge 1997).

The 1980s and 1990s represent a period of significant migration in Jamaican history—a period where economic crisis, declining income-generating opportunities and rising levels of violence led many to seek new opportunities, primarily in the United States and Canada. The crisis of the 1980s and 1990s struck hard at the social reproductive capacity of households. And, as a number of scholars have extensively documented, throughout this period, the austerity measures that indebted countries like Jamaica were obliged to implement had a significant impoverishing impact on the everyday lives and livelihoods of men and women on the island (Deere *et al.* 1990). While policies like the removal of government subsidies, higher interest rates and currency devaluations were aimed at restoring so-called "market fundamentals," for ordinary Jamaicans, these policies took the form of higher prices, fewer state protections and fewer income-generating opportunities. Women bore the heaviest burden of the structural adjustment reforms of this period, and, in particular, those who were single heads of household (a status held by 45 percent of Jamaican women in 2000) accounted collectively for two-thirds of households in poverty (World Bank 2004).

Poor women were affected not only by the rising levels of unemployment and underemployment, but also suffered from the unequal burden that the rising cost of living exacted. In addition, they incurred the loss of public services like health and education as public expenditure on these forms of social reproduction declined. For example, in 1982–3 and 1987–8, government spending on Jamaican health services in real terms declined by 71 percent, and in 1982–3 and 1986–7, spending on education declined by approximately 57 percent. Women, in particular, compensated for the loss of publicly provided social welfare by increasing the amount of the time that they devoted to the work of social reproduction. The shift in these social reproduction costs, from the state to the household and the individual, disproportionately affected women, who not only assumed the burdens of unpaid work in the home and paid work in the workplace, but also assumed the triple burden of the work needed to maintain their communities (Moser and Peake 1987; French 1994). As Peggy Antrobus (1989) concluded, the structural adjustment policies of the 1980s and 1990s were deeply embedded in a gender ideology that was fundamentally exploitative of women's time, labor and sexuality. It was a policy approach supported by international donor agencies, as well as the Jamaican government, that had little regard for the forms of economic, political and social violence that such levels of austerity generated.

Numerous studies have pointed to the importance that migration played during the 1980s and 1990s. High levels of migration, not only among the skilled middle classes but also the unskilled and poor, functioned as an important safety valve that stemmed the rate of impoverishment faced by many Jamaicans during this period. By stretching the spaces of social reproduction across international borders, Jamaicans filled the void created by the retreat of the state from social reproduction. Complex networks emerged during this period, as migrating women and men devised transnational household divisions of labor that relied upon family members and friends to maintain levels of social reproduction and upon members abroad to remit monetary payments earned abroad. Remittances grew steadily during the 1980s and 1990s, rising in real prices from US$89 million in 1980 to US$892 million in 2000 (World Bank 2003). Throughout the following decade, remittances grew steadily to a level that regularly matched the value of goods and services exported from the island.

A recent survey conducted by researchers at the University of the West Indies concurs with previous ones, that remittances have been a lifeline, which minimized the levels of impoverishment that the austerity measures of the previous two decades had generated. Because remittances often go directly to households, migrating Jamaicans were able to ensure that family members could gain access to necessary aspects of social reproduction (like education and health) that might have been otherwise unaffordable. Indeed, as the study by Thomas-Hope *et al.* (2009) found, approximately two-thirds of emigrants remitted to the household they left behind, sending back (on average) approximately US$640 annually. They also found that of remittances that were specifically earmarked for a particular use, 14 percent were allocated to education and 31 percent to child support. Another recent study by the International Office of Migration (IOM) corroborates the findings of this earlier study, citing that 54 percent of the money received by remitting households in 2011 went toward the cost of food (18 percent), utilities (19 percent), education (14 percent) and housing (9 percent) (IOM 2010).

Recognition of the remarkable consistency of remittances in the 1990s led institutions like the World Bank and the Latin American Development Bank to explore ways to bring more of these flows of capital under the oversight of states (Chami *et al.* 2003; World Bank 2003; IDB 2004; Kapur 2004; Newland 2004; OECD 2005; Orozco 2004). And by the start of the new millennium, the World Bank, the Inter-American Development Bank and the United States Agency for International Development (USAID) had already assumed a leading role in efforts to develop procedures for monitoring, measuring and maximizing the possibilities for remittances to become a form of development finance of last resort. In 2003, for example, an intra-agency task force on remittances was struck when the UK's Department for International Development (DFID) and The World Bank hosted a conference, entitled *Migrant Remittances, Development Impact, Opportunities for the Financial Sector and Future Prospects*, with over one hundred representatives of central and private banks, government policy makers, multilateral and bilateral donors, NGOs and universities from forty-two countries (Maimbo and Ratha 2003, 2005). By 2004, the subject of remittances, and their potential as a source of development finance, was on the agenda of the 2004, G8 meeting at Sea Island in Georgia (United States), signaling the importance that this form of cash transfer had assumed to both receiving and sending states.

Viewed as a type of development finance of last resort with the capacity to help countries to sustain their creditworthiness, especially during periods of economic downturn, remittances quickly transformed from a form of affective investment, crucial to the physical and psychological welfare of family members disconnected across borders (Burman 2002), to a potential pool of capital that could be channeled into national economic programs. Indeed, attention to the potential contribution of remittances to economic production culminated in a number of studies that questioned the value of remittances to Jamaica's development. One such study, by Kim (2007), concluded that remittances had a negative impact on labor participation levels because they allowed receiving individuals to heighten their reservation wage,[3] prompting much subsequent speculation over the possible benefits that taxation could generate. Another study by Bussolo and Medvedev (2007), concluded that although remittances did produce distortions that had the capacity to appreciate exchange rates, reduce country competitiveness and damage a country's export base and manufacturing sector, these transfers were still an important channel of external financing for Jamaica and an important source of income among the poor.

Between 2000 and 2010, debates regarding the impact of remittances on economic growth and productivity highlighted the devalued status of social reproduction relative to economic production among academics, government representatives and policy makers. The fact that monies directed at household consumption, education, health and housing are often considered somehow unproductive and better directed towards national investment priorities, highlights the extent to which women's efforts to keep their households out of poverty during the 1990s went unnoticed and taken for granted. For although it is a well-known fact that poverty levels in Jamaica declined substantially throughout the crisis years between the mid and late1990s, from approximately 27.5 percent of the population in 1995 to 17 percent by 1999 (World Bank 2004), the importance of this achievement to quality of life has tended not to be acknowledged by states and development agencies alike.

Thus, in their assessment of Jamaica's development trajectory, the World Bank's only comment on the role played by remittances was the observation that:

> Remittances which have grown sharply in dollar terms, though not relative to GDP, have probably kept people out of poverty, but contributed much to its decline.... A sustained reduction in poverty in the future is likely to depend on sustained growth in the Jamaican economy as well as the implementation of pro-poor policies.
>
> (World Bank 2004: 41)

While the bank identified both poverty reduction and economic growth as important components of Jamaica's future development trajectory, what emerged was an increasing orientation toward capturing migrant capital for development finance purposes. The shift in government discourse from "remitting migrants" to "an investing diaspora" has been both discursive and material; along with the changing terminology, there has been a shift in state policies toward a more exclusive incorporation of the professional, entrepreneurial and highly skilled—a group viewed as more likely to generate economic growth at the community and national levels (Kuznetsov 2006).

From remitting migrants to investing diaspora: exploring the evolution of Jamaica's diaspora option

By the late 2000s, the diaspora option had become a key element in the development arsenal of a retinue of policy makers within leading development institutions and national governments across the global South (Kunz 2008; Gamlen 2011; Pellerin and Mullings 2013). I believe that the shift in state attention from migrant to diaspora and from remittances to investment, is indicative of an ideological move away from questions of poverty, the failures of past neoliberal policies and the crisis of social reproduction, toward questions of entrepreneurship, investment and the pragmatics of diasporic networks. This shift should also be seen as both gendered and classed, to the extent that it has foreclosed any space available to address the intersecting inequalities faced by poor women within metropolitan labor markets and the responsibilities of states, in both sending and receiving countries, to respond to the exploitative conditions that many remitting migrants face. Where they have addressed questions of exploitation, policy makers have tended to focus almost exclusively on lowering the transaction costs associated with the remittance process (DIFID 2005), rather than on challenging the immigration, taxation or work environments within which migrants generate the capital to remit.

As early as 1992, the idea of engaging diasporic communities as a development strategy was considered by states in the region (West Indian Commission and Sir Shridath Ramphal 1993). The West Indian Commission report, "A time for action," advocated for a closer relationship to be forged between Caribbeans abroad and governments in the region. Noting the strength of the affective ties held by West Indians abroad for the region, as captured in their observation that "nothing could alter the fact that their navel strings were still buried in some corner of the West Indies" (West Indian Commission and Sir Shridath Ramphal 1993: 408), the commission advocated for a change in attitude on the part of the region's governments toward their expatriate populations. The call to recognition was well heeded and in 2004, the Jamaican government took its first formal step towards building a relationship with members of its diaspora by hosting a two-day conference entitled "Unleashing the potential." The conference, which attracted 250 members of the diaspora, primarily from the UK, the USA and Canada, was the first in a series of biennial conferences that continue even today.

Efforts were also made in the late 1990s to institutionalize Jamaica's diaspora strategy, establishing not only the biennial conferences but also a number of governing institutions. From the creation of an advisory board made up of diaspora representatives from the United Kingdom, the United States and Canada, and the creation of a Diaspora Foundation and Institute for the hosting of the biennial conferences, the Jamaican state has worked hard to bring a broad set of interest groups together in order to co-ordinate developmental interventions. Some of these interventions have been philanthropic in nature and have focused on social reproductive needs such as health and education, while others have been oriented toward generating trade and investment. But the key space where individual diaspora members directly communicate with each other and strategize is the biennial conference, and as such, an examination of this space can offer some insight into the ways that the state and diaspora members imagine, prioritize, and contest development. The conference also offers a window into the gender asymmetries at work within diasporic formations and their role in shaping how concerns within diaspora come to attain visibility and importance. My analysis of the transcript[4] of the 2006 biennial conference published by Franklyn (Franklyn 2010), as well as my own notes and observations of video footage of the conference, offers some indication of these asymmetries at work.

How gender asymmetries shaped the conference space can be seen in the range of voices that were represented, the range of issues addressed and the priorities given to particular concerns. Structured by the agenda generated by members of the advisory committee, the 2006 conference focused on the establishment of governance structures; the opportunities and challenges of mobilizing members of the diaspora; the crime situation; and forging business linkages. The panelists were drawn primarily from the government, the diaspora advisory council and the private sector, and the presentations were followed by a question and answer session involving all of the gathered delegates. Based on the conversations, a number of resolutions and achievable targets were tabled and voted on at the end of conference.

Although there was no explicit discussion on the social reproductive issues affecting men and women in Jamaica and the diaspora, delegates often raised concerns about the adequacy of the health and education provisions on the island, levels of crime and violence and the provisions being made for the rising number of deported Jamaicans, a figure that totaled over 27,986 between 1992 and 2006, or over 59 percent of all returned migrants (Thomas-Hope *et al.* 2009). Yet, an overarching theme throughout the conference was that of devising practical strategies to boost capital inflows to the state. Whether in the form of a bond issue, the facilitation of diaspora investors, the branding of culture and intellectual property production or diasporic investments in education and health, diaspora members engaged in discourses that reproduced the invisibility and devalued worth of the circumstances under which the vast majority of Jamaicans reproduced the space of diaspora and its associated emotions.

The invisibility of the gender relations that underpinned the feminization of migration, gender divisions of labor at home and the forms of violence (both economic and domestic) that women and men faced at home and abroad, largely reflected the range of voices at the conference and the segments of the diaspora that they represented. Given the fact that most diaspora members fund all of the costs associated with attending these conferences (i.e. airfares, accommodation and registration) themselves, it is clear that opportunities to participate are often restricted to a small minority, comprised largely of working or retired professionals, who have remained active in a variety of diaspora organizations in their home country. An examination of the transcripts from the 2006 conference also revealed that most of these voices belonged to men, and of the 128 presentations made and questions submitted, only twenty-six were posed by women.

Having more women's voices heard would not have necessarily guaranteed that the contributions of women and their experiences would have been better valued or supported, but a broader array of voices located at diverse intersecting locations of disadvantage/privilege would have potentially oriented the conversations toward a more representative range of issues.

What emerged instead were conversations that weakly gestured at some of the inequalities that positioned different members of the diaspora in ways that limited their capabilities and, consequently, their potential contribution to processes of social transformation at home and abroad. But, even where members explicitly asked questions about the asymmetries that structured diasporic practices and engagements, the responses reflected imaginaries that prioritized the needs and capabilities of men rather than women. For example, in the session on crime, a delegate from Canada raised the only question that directly addressed the situation of women in the diaspora. Responding to the Minister of National Security's claim that the unraveling of the country's social structure could be seen in the fact that only 31 percent of children live with both biological parents, the delegate's question diverted attention away from the dominant moral discourse that attributed crime to the breakdown of the family and to the politics of blame that constructed single motherhood as the root of youth marginality and crime. One participant asked the following question:

> Someone has actually spoken in terms of the gender dimension of crime, and what is happening to women, who are not only victims but some of women are also involved in criminal activities. I wonder if the Minister could speak to that because it is an important issue, particularly in the Diaspora, where women are concerned and because single mothers are sometimes blamed for what is happening to young people.
>
> (Franklyn 2010: 109)

The Minister responded:

> I noticed in a probably unintended way, or unwittingly, that gender was taken to mean woman. But let me say that the bulk of the victims and the perpetrators of violent crimes are our young men. I think that mirrors what is happening to our young men in school. … The question I ask myself is, what happened in the school system in the seven years prior to 1982 which contributed to this fall in performance of the young men? Now there are a lot of hypotheses we can look at, but I believe that one issue we can look at has to do with the men leaving the teaching professions and the absence of role models for young men. We have to address these issues frontally and we need to recognize the particular plight of the young men who see no alternative to the gun and violence.
>
> (Franklyn 2010: 112)

The Minister's response to the delegate revealed a lack of knowledge of the forms of dependency and the lack of opportunity that continues to propel many poor young women into forms of work and relationships that are both economically and domestically violent; it also revealed a prioritization of the concerns of poor young men, who in popular discourse have been constructed as an exceptionally marginalized group. The Minister's reluctance to talk about the relationship between the neoliberal strategy that the Jamaican government continues to pursue, and the socio-economic fracturings, dislocations and transnational circuitries of sadness that these strategies have produced, serves to highlight the power that the logic of the market continues to hold in state imaginaries of diaspora-led futures. A similar response was given when another delegate stated:

> We have been given some explanation for the underlying social issues that are behind the crimes. I would like to know, what is your policy agenda for tackling the social-economic problems?
>
> (Franklyn 2010: 110)

Responding to this question, the Minister stated:

> We do have within the Ministry a community security initiatives (sic) and a social justice programme which focuses on mediation programmes and the like, but I think the larger range of social interventions are going to have to fall to other ministries. . . . We cannot afford another generation to come out of the school system uncertified and without the capacity for gainful employment and health participation in society. The other critical dimension is how do you facilitate the small, and indeed the micro enterprises which will generate your greatest impact on employment possibilities in the near term?
>
> (Franklyn 2010: 110–11)

The invisibility of the importance to national development of the paid and unpaid labor of women at home and abroad, and the need to support their efforts to challenge the systems of oppression that affect them, can be seen in the resolutions that were finally adopted at the 2006 conference. At this final session, a resolution was put forward that stated:

> We propose that the Jamaican Diaspora adopt as its core programme thematic issues, for example, trade and economic development; security (criminal and civil justice systems); health and social issues; immigration; education; culture and sports; women's issues and churches/faith.
>
> (Franklyn 2010: 293)

But, this proposal was amended when one of the delegates, offering a friendly amendment, suggested that the term "women's issues" should be replaced by the term "gender issues." Like the Minister of National Security, whose earlier comments suggested that speaking about the situation of women might deflect attention away from the plight of poor young men, this delegate also expressed his concern that men might be excluded from the attentions and interventions of the diaspora if specific attention were to be called to issues affecting women. As this delegate argued:

> I do not know if you would be prepared to accept what I regard as a friendly amendment. Instead of "women's issues," "gender issues," because a word is critical, but also I think we would be fooling ourselves not to also consider that there are many males at risk and I think if you want to keep "women," you should also add that factor of gender which is more all inclusive and I think it is much more all encompassing, an amendment that the delegates agreed with.
>
> (Franklyn 2010: 293)

The idea that identifying the condition of women would somehow exclude concerns for the plight of poor men is indicative of the importance of context to the politics of naming. Ultimately, what appeared to be an attempt to open up the field of diasporic intervention to the range of issues facing both men and women located at the intersection of different systems of oppression, ended up reproducing the popular discourse of the marginalized Jamaican male—a discourse that ultimately crowded out any attempt to recognize the issues affecting women's lives and their contributions to the strength of the diaspora. Therefore, it is not surprising that, in the end, the only gender issues discussed by the delegates at the conference were, in fact, those related to the marginalization of young men, crime and violence.

Conclusion

The framing of the diaspora option as a straightforward solution to the problem of competitiveness and current efforts to build partnerships between states and migrants, are part of a broader neoliberal strategy aimed at creating a new middle class in the global South, with individual and collective interests that are not only more closely oriented toward the norms of the market but also to the social, political and economic interests of those who are dominant within it (Mullings 2012; Bakewell 2008; Larner 2007). But, with their prioritization of diaspora members who are positioned to make contributions to national development finance objectives, those touting the diaspora option are in danger of rendering invisible the value of the contributions and conditions under which poor women, many of whom are concentrated in the lowest paid and marginal segments of labor markets abroad, produce diaspora. It is ironic that the socio-economic fracturings, dislocations and migratory flows that have emerged from the liberalization of markets over the last thirty years are the same processes that continue to generate the transnational circuitries of sadness, diasporic yearning and the faithful flows of capital that exist at the heart of the diaspora option that governments seek to institutionalize. Deborah Thomas argues that as much as diaspora might be imagined as a site of political aspiration and solidarity, and as a social, cultural, and political rubric (Campt and Thomas 2008; Thomas 2008), it can also be seen as a historically contingent strategy to advance particular interests that reflect the hegemony of asymmetries across race, class, gender and sexuality, which structure diaspora formations both internally and externally. The current popularity of the diaspora option seems simple and pragmatic, because it focuses on members of a country's diaspora who are imagined to be the most skilled, the most entrepreneurial and the most networked. But, diaspora formations are not solely the preserve of the highly skilled and globally networked; they are often also forged from emotional circuits linked to semi-involuntary migration, trauma, violence and sadness—all of which reflect the ways that race and gender shape the circulation of global capital and the roles that Jamaican men and women have come to play within them. Without a feminist understanding of the ways that diaspora formations are linked to intersecting systems of oppression, the contributions of its most marginalized members will continue to be rendered invisible to the detriment of the sustainability of diaspora itself.

Notes

1 This work was carried out with the aid of a grant from the Social Science Research Council of Canada (SSHRC), Ottawa, Canada. I would like to thank Ibipo Johnston-Anumonwo and Ann Oberhauser for their support and patience in the writing of the paper, and also two anonymous reviewers who took the time to provide incredibly helpful comments that have enormously enriched this paper. Any unintentional errors and omissions, of course, remain entirely mine.
2 I draw here on the work of Foucault (1991), who uses the term "technology of power" to refer to the practices, strategies and techniques used by a variety of institutions (e.g. states, families, religious organizations) to shape the conduct of a given population in order to produce a desired effect.
3 A reservation wage refers to the lowest wage rate that a worker would be willing to accept for a particular type of job.
4 The text that appears in Franklyn's book (Franklyn 2010) is not a complete transcript of the dialogue that took place at the conference. There are some comments raised by persons in attendance that are not in the book, but the text nevertheless provides a reliable reproduction of the conversations that took place.

References

Antrobus, P. (1989) "Crisis, challenge and the experiences of Caribbean women," *Caribbean Quarterly* 35: 17–28.

Bakewell, O. (2008) "Keeping them in their place: the ambivalent relationship between development and migration in Africa," *Third World Quarterly* 29: 1341–58.

Bauer, E. and Thompson, P. (2006) *Jamaican Hands Across the Atlantic*, Kingston: Ian Randle Publishers.

Brubaker, R. (2005) "The 'diaspora' diaspora," *Ethnic and Racial Studies* 28(1): 1–19.

Burman, J. (2002) "Remittance: or diasporic economies of yearning," *Small Axe* 2: 49–71.

Bussolo, M. and Medvedev, D. (2007) "Do remittances have a flip side? A general equilibrium analysis of remittances, labour supply responses and policy options for Jamaica," World Bank Policy Research Working Paper No. 4143, Washington, DC: World Bank.

Campt, T. and Thomas, D. A. (2008) "Gendering diaspora: transnational feminism, diaspora and its hegemonies," *Feminist Review* 90: 1–8.

Chami, R., Fullenkamp, C. and Jahjah, S. (2003) "Are immigrant remittance flows a source of capital for development?," International Monetary Fund (IMF) Working Paper No. 03/189, Washington, DC: International Monetary Fund.

Deere, C., Antrobus, P., Bolles, L., Meléndez, E., Phillips, P., Rivera, M. and Safa, H. (1990) *In the Shadows of the Sun: Caribbean Development Alternatives and US Policy*, Oxford: Westview Press.

Department for International Development (DFID) (2005) "Sending money home?: a survey of remittances, products and services in the United Kingdom," DFID Report, London: Department for International Development.

Ehrenreich, B. and Hochschild, A. (2002) *Global Woman: Nannies, Maids and Sex Workers in the New Economy*, New York: Henry Holt.

Elson, D. and Pearson, R. (1981) "The subordination of women and the internationalisation of factory production," in K. Young, C. Wolkowitz and R. McCullagh (eds) *Of Marriage and Market*, London: Routledge and Kegan Paul, pp. 144–66.

Erel, U. (2012) "Introduction: transnational care in Europe—changing formations of citizenship, family, and generation," *Social Policy* 19: 1–14.

Fog Olwig, K. (1999) "Narratives of the children left behind: home and identity in globalised Caribbean families," *Journal of Ethnic and Migration Studies* 25: 267–84.

Fog Olwig, K. (2007) *Caribbean Journeys: An Ethnography of Migration and Home in Three Family Networks*, Durham, NC: Duke University Press.

Foucault, M. (1991) "Governmentality," (lecture at the Collège de France, February 1, 1978), in G. Burchell, C. Gordon and P. Miller (eds) *The Foucault Effect: Studies in Governmentality*, Hemel Hempstead: Harvester Wheatsheaf, pp. 87–104.

Franklyn, D. (ed.) (2010) *The Jamaican Diaspora: Building an Operational Framework*, Jamaica: Wilson Franklyn Barnes.

French, J. (1994) "Hitting where it hurts most: Jamaican women's livelihoods in crisis," in Pamela Sparr (ed.) *Mortgaging Women's Lives*, London and New Jersey: Zed Books.

Fudge, J. (1997) "Little victories and big defeats," in A. A. B. Bakan and D. K. Stasiulis (eds) *Not One of the Family: Foreign Domestic Workers in Canada*, Toronto: University of Toronto Press, pp. 119–46.

Gamlen, A. (2011) "Creating and destroying diaspora strategies," IMI Working Paper No. 31, International Migration Institute, University of Oxford.

Glick Schiller, N. and Fouron, G. E. (2001) *Georges Woke up Laughing: Long Distance Nationalism and the Search for Home*, Durham, NC: Duke University Press.

Inter-American development Bank (IDB) (2004) *Sending Money Home: Remittance to Latin America and the Caribbean*, Washington, DC: Inter-American Development Bank.

International Office of Migration (IOM) (2007) Jamaica Mapping Exercise, London: International Office of Migration.

International Office of Migration (IOM) (2010) *Migration in Jamaica: A Country Profile 2010*, Geneva, Switzerland: International Office of Migration.

Ionescu, D. (2006) *Engaging Diasporas as Development Partners for Home and Destination Countries: Challenges for Policymakers—MRS No 26* in *MRS No 26*—Geneva: International Organization for Migration.

Itzigsohn, J., Cabral, C. D., Medina, E. H. and Vázquez, O. (1999) "Mapping Dominican transnationalism: narrow and broad transnational practices," *Ethnic and Racial Studies* 22: 316–39.

Kapur, D. (2004) *Remittances: The New Development Mantra?*, Geneva: UNCTD.

Katz, C. (2001) "Vagabond capitalism and the necessity of social reproduction," *Antipode* 33(4): 709–28.

Kim, N. (2007) "The impact of remittances on labor supply: the case of Jamaica WPS4120," World Bank Policy Research Working Paper No. 4120, February 2007: 1–18.

Kofman, E. (2012) "Rethinking care through social reproduction: articulating circuits of migration," *Social Politics* 19: 142–62.

Kunz, R. (2008) "Mobilising diasporas: a governmentality analysis of the case of Mexico," Working Paper Series "Glocal Governance and Democracy 03," Institute of Political Science, Lucerne: Switzerland, University of Lucerne. Available online at: www.unilu.ch/files/Diaspora-governing_-wp03.pdf (accessed June 11, 2013).

Kuznetsov, Y. (2006) *Diaspora Networks and the International Migration of Skills: How Countries can Draw on their Talent Abroad*, Washington, DC: World Bank.

Larner, W. (2007) "Expatriate experts and globalising governmentalities: the New Zealand diaspora strategy," *Transactions of the Institute of British Geographers* 32: 331–45.

Lewis, W. A. (1954) "Economic development with unlimited supplies of labour," *Manchester School of Economics and Social Studies* 22: 139–91.

Lowenthal, D. (1972) *West Indian Societies*, Oxford: Oxford University Press.

McDowell, L. (2009) *Working Bodies: Interactive Service Employment and Workplace Identities*, Oxford: Wiley-Blackwell.

Maher, K. H. (2004) "Globalized social reproduction: women migrants and the citizenship gap," in A. Brysk and G. Shabir (eds) *People out of Place: Globalization, Human Rights and the Citizenship Gap*, New York: Routledge, pp. 131–52.

Maimbo, S. M. and Ratha, D. (2003) *Remittances: Development Impact and Future Prospects*, Washington, DC: World Bank.

Maimbo, S. M. and Ratha, D. (2005) *Remittances: Development Impact and Future Prospects*, Washington, DC: World Bank.

Momsen, J. (1992) "Gender selectivity in Caribbean migration," in S. Chant (ed.) *Gender and Migration in Developing Countries*, London: Belhaven Press, pp. 73–90.

Moser, C. and Peake, L. (1987) *Women, Human Settlements and Housing*, London: Tavistock Publications.

Mullings, B. (2012) "Governmentality, diaspora assemblages and the ongoing challenge of 'development'," *Antipode* 44: 406–27.

Neumayer, E. (2006) "Unequal access to foreign spaces: how states use visa restrictions to regulate mobility in a globalized world," *Transactions Institute of British Geographers NS* 31: 72–84.

Newland, K. (2004) *Beyond Remittances: The Role of Diaspora in Poverty Reduction in their Countries of Origin*, Washington, DC: Migration Policy Institute for the Department of International Development.

OECD (2005) *The Development Dimension Migration, Remittances and Development*, Paris: OECD.

Orozco, M. (2004) "Remittances to Latin America and the Caribbean: Issues and Perspectives on Development, 2," Washington, DC: Report Commissioned by the Organization of the American States.

Pellerin, H. and Mullings, B. (2013) "The 'Diaspora Option,' migration and the changing political economy of development," *Review of International Political Economy* 20: 89–120.

Pessar, P. R. and Mahler, S. J. (2003) "Transnational migration: bringing gender," *International Migration Review* 37: 812–46.

Portes, A. (1995) *The Economic Sociology of Immigration: Essays on Networks, Ethnicity, and Entrepreneurship*, New York: Russell Sage Foundation.

Pratt, G. (2009) "Circulating sadness: witnessing Filipina mothers' stories of family separation," *Gender, Place and Culture—A Journal of Feminist Geography* 16: 3–22.

Rodriguez, R. M. (2010) *Migrants for Export: How the Philippine State Brokers Labor to the World*, Minneapolis, MN: University of Minnesota Press.

Safa, H. (1995) *The Myth of the Male Breadwinner: Women and Industrialization in the Caribbean*, Boulder, CO: Westview Press.

Safran, W. (1991) "Diasporas in modern societies: myths of homeland and return," *Diaspora: A Journal of Transnational Studies* 1: 83–99.

Sheffer, G. (2006) *Diaspora Politics: At Home Abroad*, Cambridge: Cambridge University Press.

Shuval, J. T. (2000) "Diaspora migration: definitional ambiguities and a theoretical paradigm," *International Migration* 38(5): 41–56.

Silvey, R. and Lawson, V. A. (1999) "Placing the migrant," *Annals of the Association of American Geographers* 89: 121–32.

Sives, A. (2012) "Formalizing diaspora-state relations: processes and tensions in the Jamaican case," *International Migration* 50(1): 113–28.

Smith, M. P. (2000) *Transnational Urbanism: Locating Globalization*, Oxford: Blackwells Publishers.

Thomas, D. A. (2008) "Wal-mart, 'katrina,' and other ideological tricks: Jamaican hotel workers in Michigan," *Feminist Review* 90: 68–86.

Thomas-Hope, E. (1999) "Return migration to Jamaica and its development potential," *International Migration* 37: 183–207.

Thomas-Hope, E., Kirton, C., Knight, P., Mortley, N., Urquhart, M. A., Noel, C., Robertson-Hickling, H. and Williams, E. (2009) *Development on the Move: Measuring and Optimising Migration's Economic and Social Impacts: A Study of Migration's Impacts on Development in Jamaica and How Policy Might Respond*, Delhi and London: Global Development Network and Institute for Public Policy Research.

Trotz, D. A. (2006) "Rethinking Caribbean transnational connections: conceptual itineraries," *Global Networks* 6: 41–59.

Trotz, D. A. (2007) "Going global?: transnationality, women/gender studies and lessons from the Caribbean," *Caribbean Review of Gender Studies: A Journal of Caribbean Perspectives on Gender and Feminism* 1: 1–18.

United Nations (2006) "Feminization of migration remittances, migrants' rights, brain drain among issues, as population commission concludes debate," in UN Economic and Social Council POP/945 Commission on Population and Development Thirty-ninth Session 5th and 6th Meetings (AM and PM), New York, United Nations.

West Indian Commission and Sir Shridath Ramphal (1993) *Time for Action: Report of the West Indian commission*, Kingston: University of the West Indies Press.

World Bank (2003) "International migration, remittances and the brain drain. A study of 24 labor-exporting countries", Policy Research Working Paper No. 3069, Washington, DC: World Bank.

World Bank (2004) "The road to sustained growth in Jamaica: A World Bank country study report," Washington, DC: World Bank.

World Bank (2007) "Concept note: mobilizing the African Diaspora for development," Washington, DC: Capacity Development Management Action Plan Unit (AFTCD), Operational Quality and Knowledge Services Department (AFTQK).

Wright, M. W. (2006) *Disposable Women and Other Myths of Global Capitalism*, New York: Routledge.

4 Stuck in a groove?

Gender, politics and globalization in anti-sex trafficking policy initiatives

Vidyamali Samarasinghe

Introduction

Across the globe, female sex trafficking is embedded in a complex network of source countries, transit points and destination countries that rely upon the exploitation of sex trafficking victims, many of whom are women and children. This sex trafficking network overlaps with and takes advantage of the often ignored and inconsistent migration and immigration policies and practices in these countries. Responding to the three "Ps" (Prevention of trafficking, Protection of victims and Prosecution of traffickers)[1] identified by the United Nations Convention Against Transnational Organized Crime and its Protocol to Prevent, Suppress, and Punish Trafficking in Persons, Especially Women and Children in 2000[2] (hereafter referred to as the UN Protocol), many member states have adopted anti-trafficking policies. These measures represent an attempt to stem the tide of sex trafficking in an era of unprecedented cross-border female migration, which has been triggered by the current forces of economic and political globalization. Despite these efforts, sex trafficking often hides behind this increase in female migration as it makes victims harder to identify and criminal action against traffickers more difficult to enforce.

In general, countries of the global North are the major destinations of sex trafficking and countries of the global South are the major source regions for this trafficking. However, contiguous land borders between richer and poorer neighbors of the global South remain porous and often blur the distinctions between what is designated as source, transit and/or destination countries. With the introduction of the UN Protocol, a body of literature has emerged on anti-trafficking policy initiatives, which analyzes different dimensions of cross-border female sex trafficking (see Bales and Lise 2005; Doezema 2005, 2010; Jeffreys 2009; Kempadoo 2005; Samarasinghe 2008; Shelley 2010). In this chapter, I examine some of the challenges faced by both local and global actors in designing and implementing cross-border, anti-sex trafficking policies.

While the accuracy of numbers of females trafficked into the commercial sex sector is often questioned, there is common agreement that cross-border sex trafficking does occur and accounts for a significant proportion of global human trafficking (US Department of State (USDOS) 2012). The anti-trafficking policy guidelines offered by the UN Protocol are firmly focused on "victim centered" initiatives. This chapter scrutinizes the challenges in identifying female "victims" in the sex trafficking policy discourse. The contention of this study is that female "victims" of sex trafficking are literally caught in the middle of two cross-cutting issues, i.e. transnational migration regimes and ideologically driven political debate on prostitution and sex work. These two issues have a decisive impact on the formulation of effective anti-sex trafficking measures capable of transcending sovereign borders of an otherwise highly interconnected, globalized world.

First, cross-border sex trafficking, by definition, falls within international parameters; thus, anti-sex trafficking policies often face the problem of incompatibility with the existing international migration policies of sovereign states.[3] Governments are guided by practical self-interest based on specific socio-economic and political priorities in relation to migration issues, and they often prioritize the enforcement of laws pertaining to international migration over anti-sex trafficking initiatives. Women and girls who become victims of sex trafficking are often reduced to de-personalized bodies with no voice at either end of transnational migration flows, i.e. source countries and destination countries. They face a double jeopardy as victims of sex trafficking as well as (often) illegal immigrant status.

In addition, the "ideological brawl that has for decades dominated all discussions of prostitution" (Beloso 2012: 48) has driven the anti-trafficking policy initiatives among member states in different directions, making it almost impossible for the emergence of practical, cohesive anti-sex trafficking policy initiatives at the global level. Scholars who strongly ground themselves on the abolitionist platform of radical feminism argue that female prostitution is a violation of the human rights of women as well as an exploitative practice of patriarchal societies (Jeffreys 2009; Raymond 2004). At the opposite end of the spectrum, feminists who advocate for the right of women to engage in sex as work also vehemently argue that, as with many other labor activities, the exploitation of women in the sex trade should be defined as another form of labor exploitation; they also suggest that stigmatizing sex work by females results from male hegemony (Kempadoo 2005). While this debate showcases a major challenge for feminists "to achieve solidarity across difference, because there is no simple 'us' in feminism" (Cornwall *et al.* 2007: 3), the inability to arrive at a consensus on the issue of cross-border sex trafficking stems from fundamentally opposing views on the very definition of the concept of "victim" in sex trafficking.

This chapter is divided into four main sections. Following the introduction, the second section examines policy ramifications stemming from situating cross-border female sex trafficking with transnational migration as it pertains to both source countries and host countries of female sex trafficking. The third section is devoted to an analysis of how anti-sex trafficking policy is impacted by contrasting ideological stands on the female commercial sex industry. In the concluding section, I bring attention to the stalled policy discourse in anti-sex trafficking initiatives in relation to the "victim centered" approach identified by the UN Protocol.

Remittances versus cheap labor: The north/south divide?

The global South: Source countries

In the case of poorer source countries, the socio-economic and political reality of contemporary transnational migration is often informed by the impact of globalizing forces. As Kapur (2003) explains, the ubiquitous process of the free flow of capital is deemed critical to the efficiency of the market and intrinsic to the globalization process; however, the same market mechanisms also trigger a global flow of labor. One of the most significant outcomes of this labor mobility is the "feminization" of migration and work, whereby new global spaces are opened for poor, less-educated women (from lower-income countries) seeking employment across borders. The same global flow of labor that traps women and girls into sex trafficking has prompted researchers to identify sex trafficking as the "underside" or the "darker side" of globalization (Sassen 2000; Samarasinghe 2003).

Given the economic demand for labor migration, largely from low- to high-income countries, how could poorer countries formulate policies to prevent sex trafficking? *Should* poorer countries, which are source regions of cross-border sex trafficking, formulate migration policies that could identify females who may enter sex trafficking from the thousands of women who migrate across borders to seek work overseas? From a practical economic standpoint, why would poorer source countries undertake such a time-consuming and expensive venture, since migrating women often decrease unemployment and underemployment levels in the source country and bring a significant cache of much needed foreign currency earnings as remittances? Consequently, poorer countries will thus have limited interest in controlling outward migration, be it legal or illegal (Kapur 2008: 113).

As long as remittances remain a significant source of income from overseas employment, methods of separating sex trafficking of women from other employment-related migration seem to get minimal attention from government policy makers of immigration source countries. Experts anticipate that remittance flows will continue growing, with global levels of revenue from remittances expected to reach $615 billion by 2014, of which $467 billion will flow into developing countries (Ratha *et al.* 2012). While women account for 43.1 percent of international migrants in the age group between twenty and sixty-four years, in some countries (such as the Philippines), female migrants overseas account for 51.1 percent of total transnational migrants (ibid.). The notable increase in a variety of cross-border circuits, which are sources of livelihood, profit making and accrual of foreign currency earnings, conceptualized as the "counter geographies of globalization," include the illegal trafficking of people for the sex industry (Sassen 2000: 504).

The Philippines is one of the first developing countries to undertake a proactive policy of using labor exports as a strategy of development, specifically female migration.[4] The case of Filipina entertainers trafficked to Japan demonstrates how a source country explicitly encourages female migration for a specific type of employment in order to secure foreign currency earnings via remittances, thereby easing its unemployment problem (Samarasinghe 2008). However, this strategy poses a tremendous risk to the young women who avail themselves of this opportunity. This concern is based on mounting evidence that many of the young women who choose to go overseas as "entertainers" are trafficked into the sex industry in Japan and Korea (Cameron 2008; ILO 2005; Kondo 2011; Shelley 2010; UNODC 2003).

In the course of my fieldwork on female sex trafficking in Nepal, Cambodia and the Philippines, NGO personnel who worked in anti-trafficking organizations, especially in the Philippines, were quick to point out that earning an income was the primary motive for travel overseas for young females (Samarasinghe 2008). They also indicated that while it has enacted legislation to stem the tide of sex trafficking of Filipinas who travel overseas to become entertainers, the government also depends on the remittances that overseas female entertainers send back to the Philippines. As one NGO worker in Quezon City commented, "being too strict about sex trafficking will also frighten off young women from looking for work overseas and thus depriving the foreign currency earnings for the government. Go figure!" As discussed earlier, this comment underscores some of the tensions in the discourse on enacting anti-sex trafficking legislation when the country encourages female labor export.

In addition, many of the female sex trafficking victims from poor countries of the global South are likely to be undocumented. Corruption among immigration and law enforcement personnel who are in the pockets of traffickers is known to facilitate this illegal migration of young women who are later trafficked into the sex industry overseas (Zhang 2007; Samarasinghe 2008).[5] However, it is unclear how much priority governments will give to actively pursuing anti-sex trafficking measures. Many women who are caught in the trafficking trap tend to come from poorer, marginalized classes and therefore do not usually command much attention from the male-dominated political elite. Furthermore, imposing stringent restrictions on cohorts of women who are identified as "at-risk" for sex trafficking would also hinder the transfer of remittances, which account for a significant segment of the country's foreign currency earnings.

The global North: Destination countries

While the poorer source countries are at best mishandling emigration issues in relation to anti-female sex trafficking policies, richer, demand-centered destination countries emphasize immigration issues in trafficking-related policy and law enforcement practices. Although they need the cheap labor provided by poorer countries, these destination countries demand that migrants follow the immigration laws of the host country and arrive with proper visas, authorized by their consular divisions for a select set of legal employment opportunities that require particular types of skills. Holding immigrants accountable to the legal requirements of the host nation is also a means for wealthier countries to maintain their own socio-political and economic advantage by placing the blame for criminal/undocumented migration on immigrants (Hesford and Kozol 2005; Hua 2011).

Migration policies of host countries resort to apprehension, incarceration and deportation of undocumented (often unskilled) migrants, despite the UN Protocol's emphasis on treating trafficked persons as "victims" of a crime rather than as "criminals" who have violated the immigration laws of the country (Scarpa 2008). To institute protection measures in anti-trafficking policies, the victimhood of the person has to first be established. This poses an immediate dilemma for the law enforcement of the destination country where she is often held. If she has entered the country as an illegal immigrant, the state has the legal right to take her into custody for violating the immigration laws of the country. In countries where prostitution is illegal, the immediate response of law enforcement is to charge women, who are also undocumented workers, on both counts. The primary mechanism for dealing with undocumented immigrants in destination countries, including trafficked persons, has been deportation (Apap 2003).

Although many countries across the globe have laws against female sex trafficking, it is very difficult and time consuming to establish whether or not a person is indeed a victim of trafficking. While governments are expected to look beyond the issue of illegal immigration, which is the first offence committed by a trafficked victim, very few border patrol and law enforcement officers are specifically trained to identify victims of trafficking. In the Netherlands (where prostitution is legal), the fight against illegal immigration by the state conflicted with the protection of the human rights of the victim when addressing the "cleanup" operation of streetwalkers in Amsterdam and Rotterdam. Many of those swept up during this operation were immediately deported to their source country of Bulgaria without following the procedure outlined in the victim identification of sex trafficking protocol (Hopkins and Nijboer 2004). Cross-border trafficking such as this complicates the issue of jurisdiction. In Japan, where sex trafficking of Philippine, Korean, Thai, Colombian, Ukrainian and Russian women is linked to the thriving sex industry, foreign prostitutes apprehended in a brothel raid receive little sympathy from the police (Human Rights Watch 2000). Punitive treatment is applied, regardless of the conditions under which the women migrate to (and work in) Japan, because this is seen as an immigration issue rather than a trafficking issue.

In an effort to facilitate the prosecution of traffickers, and to ensure at least temporary protection for victims, many destination countries such as the US, the Netherlands, Belgium and Italy offer temporary visas to trafficked victims who are also illegal immigrants. The "T-visa," or "continued presence provision" in the US, is provided for those who decide to testify against their traffickers. However, Bales and Lise show that the provision of the T-visa is "inconsistent and fraught with challenges," where highly stringent evidence requirements and difficult filing procedures have been cited as further obstacles (Bales and Lise 2005: 17). Certain destination countries, such as Belgium and the Netherlands, offer a short-term "reflection delay," so that a victim may decide whether or not they want to testify against the trafficker. These provisions serve mainly to support prosecution cases against traffickers. In the event that she does not want to testify, the victim is not entitled to a temporary residency visa and is likely to be repatriated. Pearson (2002) notes that those countries that provide the reflection delay are more likely to press charges against the traffickers than countries that do not have this provision. However, during the "cleanup" operation of street prostitutes in Amsterdam and Rotterdam in October 2003, none of the women who were detained were offered a reflection delay period (Hopkins and Nijboer 2004).

Another migration-related problem in terms of female sex trafficking, is the legal issue of *non-refoulement*. The principle of *non-refoulement* is a well-established international law that prohibits sending an individual back to a country where she or he is likely to be persecuted or will suffer ill treatment through torture, inhuman and degrading treatment or punishment. Anti-trafficking advocates seeking protection for repatriated victims urge the use of the principle of *non-refoulement* in order to ensure that victims do not get persecuted after repatriation. However, there are no serious efforts to harmonize laws to ensure *non-refoulement* between destination and source countries. Indeed, many sex trafficked victims may run the risk of being simply deported, or repatriated to the previous point of origin, with no support or assistance upon arrival.

Thus, poorer source countries and richer destination countries seem to have almost diametrically opposing interests with regard to the issue of migration. Consequently, despite the ratification of the UN Protocol on Trafficking and Migrant Worker Convention by primary source, transit or destination countries of sex trafficked women and girls, the lack of convergence of migration-related policies on sex trafficking between supply-based source countries and demand-based destination countries remains a serious problem.

We are all feminists: Locating prostitution in anti-sex trafficking policy initiatives

One of the most powerful and divisive factors in policy formulation on combating female sex trafficking is the issue of female prostitution. Protagonists on both sides of the debate ground their arguments on the basis of different interpretations of feminism. This contentious debate has also had a significant impact on the design and implementation of anti-sex trafficking measures across the globe, and on cross-border female sex trafficking. Prostitution regimes adopted by different governments are loosely categorized as abolitionist, prohibitionist, legalized and regulatory (Outschoorn 2004; Agustin 2007). Each of these categories encapsulates a principal ideological stand on prostitution, which in general is a variation of the abolitionist and the free choice stands on the discourse.

Abolitionist perspective on prostitution, sex trafficking and demand

In the current debate on female sex trafficking, the Coalition against Trafficking of Women (CATW) mainly spearheads the abolitionist perspective on prostitution. The abolitionist perspective, identified with radical feminism, views prostitution as sexual slavery and trafficking as an intrinsic component of prostitution (Raymond 2004). The abolitionist lobby is based on the conviction that female prostitution is inherently harmful to women, and it contends that female prostitution is a manifestation of male violence against women. According to this perspective, since sex trafficking and prostitution are synonymous, female prostitution must be eliminated in order to eliminate sex trafficking (Barry 1995; Jeffreys 2009; Raymond 2004). The CATW specifically notes that all prostitution exploits women, regardless of women's consent, and that local and global sex industries are systematically violating women's rights on an ever-increasing scale (CATW 1998).

According to this perspective, the solution is to decriminalize women and children involved in prostitution and criminalize the buyers and anyone else who promotes sexual exploitation—particularly pimps, recruiters and traffickers. These abolitionists take a proactive part in urging individual states to design and implement measures to punish the consumers of prostitution. Thus, the demand side of sex trafficking is explicitly identified, and abolitionists strongly urge countries to criminalize perpetrators. Moreover, the female prostitute who constitutes the supply side is deemed the exploited individual, who should not be punished.

Sale of sex as a form of work

The pro-free choice sex work lobby (hitherto referred to as free choice) challenges the abolitionist view that all prostitution is exploitative or harmful to women. The major proponent of this view, the Global Alliance Against Trafficking of Women (GAATW), states that although sex trafficking should be eliminated, female sex trafficking should be separated from free choice sex work, which should be legalized. This advocacy group contends that some adult women make informed choices about becoming prostitutes as a way of employment, and they should have the right to do so (Doezema 2005, 2010; GAATW 1994; Bernstein 2010; Wijers and Lap-Chew 1997). The pro-free choice prostitution advocates document trafficking not as the enslavement of women but as *"the trade and exploitation of labor under conditions of coercion and force"* (Kempadoo 2005: viii. Italics in the original).

As far as the free choice sex work lobby is concerned, "it is the lack of protection for workers in the sex industry, rather than the existence of a market for commercial sex in itself, that leaves room for extremes of exploitation, including trafficking" (Anderson and Davidson 2004: 16). Legal, free choice sex workers depend on a vibrant legal sex market to sustain their work and livelihoods. As readily acknowledged by the pro-free choice sex work protagonists, female sexual exploitation does occur in the commercial sex industry. In particular, it is argued that policy measures to criminalize paying for sex would also mean that sex workers have no legal clients, compounding the much-documented stigma and marginalization faced by sex workers, and creating an even more illicit and informal industry (Sanders and Campbell 2008).

Impact on state policies

Many experts pose questions about how these countries negotiate existing laws on prostitution/sex work and initiate anti-sex trafficking policy measures. Since the passage of the UN Protocol, a few wealthy destination countries have taken an explicit stand on the legalization of prostitution and the separation of sex trafficking from this issue. Hopkins and Nijboer (2004) claim that the Netherlands has chosen a pragmatic approach to sex trafficking, where the main focus is on regulating the prostitution business. The Trafficking in Persons Bill, passed into law in the Netherlands, made a distinction between forced and voluntary prostitution, with provision for legal penalties for those charged with trafficking. Municipalities are given the responsibility of setting up safety and health standards within the framework of the bill (Outschoorn 2004). Germany has a similar law on prostitution.

Contrasting policies among neighboring countries in Europe provide further insight into the limitations of country-specific laws in combating an activity that has a global reach. In 1999, Sweden introduced new legislation to criminalize the purchase of commercial sex, while decriminalizing prostitution itself. The policy on criminalization of the purchase of prostitution in Sweden was meant to destroy the market for commercial sex. Svanstrom (2004) explains that there was more or less unanimous support among the feminists in the established political parties in Sweden for seeing prostitution as the patriarchal oppression of women. This model employs a law enforcement strategy to apprehend and charge the client and not to punish the female prostitute, who is explicitly identified as the more vulnerable partner exploited by the dominant male. In neighboring Finland, while prostitution is not forbidden, profiteering from someone else's prostitution is criminalized (Di Nicole 2004). Finland's law is mainly aimed at clients, and prostitutes are not criminalized. However, with the introduction of the abolitionist law, Finnish customers reportedly travel overseas, especially to the "East," i.e. Russia, the Baltic States and the Far East, in search of commercial sex (Martilla 2003).

The lack of harmonization of the fundamental policy issues on the legality/illegality of prostitution among neighboring states could lead to a new movement of migration-led sex trafficking to those countries where prostitution is legal. Criminal groups are quick to grasp the opportunity to move females from Africa and newly independent states of Europe to countries where prostitution is now legal, in order to meet the increased demand fostered by the criminalization of clients in neighboring states. In particular, Shelley notes that neither criminalization of clients nor legalization of prostitution has eliminated the problem of sex trafficking. Furthermore, Dutch policies have intensified the organized crime component of prostitution markets (Shelley 2010). Meanwhile, the female "victim" remains "stuck in the groove" of vulnerability and exploitation.

The US Trafficking Victims Protection Act of 2000 (TVPA), a major milestone in anti-trafficking policy in the US, has taken an explicit abolitionist position with regard to prostitution and sex trafficking abroad, and a prohibitionist perspective on domestic prostitution. The US government has clearly adopted a strong position against prostitution, noting that "prostitution is harmful and dehumanizing and fuels trafficking in persons" (USDOS 2008: 23). This position has had a remarkable influence on designing and implementing anti-trafficking policy initiatives, especially among donor-dependent countries. As Cheung (2006) notes, the US has assumed the role of a Global Sherriff by using aid conditionality to impose its abolitionist perspective on aid-dependent, poorer countries in the global South.

Japan is a known destination for high-volume sex trafficking from Asia, as well as Eastern European and Latin American countries (Cameron 2008; Human Rights Watch 2000; ILO 2005). The Japanese sex industry is estimated to account for 1 to 3 percent of Japan's gross national product (GNP), a share equal to the Japanese defense budget. Anderson and Davidson (2004) observe that despite the prevalence of a vibrant sex industry, highly educated Japanese customers of commercial sex are least likely to be aware of the trafficking of women and children in the sex trade. It is the only country among industrialized nations placed in the Tier 2 category in the US Trafficking in Persons Reports (TIP), that does not fully comply with the minimum requirement to curb trafficking (USDOS 2012).[6] Prostitution was outlawed in Japan in 1956, but illegality has had a minimal effect on the Japanese sex industry. Moreover, Japanese policy makers have not made any concerted efforts to effectively combat the sex trafficking that feeds the sex industry in Japan.

Decrease demand or legalize prostitution?

As demonstrated in this discussion, some countries take an abolitionist perspective and attempt to target demand by criminalizing the buyer and protecting the prostitute. Other countries have defined free choice sex work as a form of legitimate labor in order to protect the interests of the woman and her right to work, while also initiating laws to punish traffickers (primarily). However, criminalization of the buyer has not resulted in any noteworthy decrease in demand for prostitution or sex trafficking. According to the abolitionist perspective, the problem of gender subordination and the resultant sexual exploitation of females by males are systemic. Also, there is an implicit portrayal of these women as helpless "victims," who lack any agency.

In contrast to the abolitionist perspective, the pro-free choice, prostitution-based anti-trafficking policy seems to view the trafficked woman as a residual factor. Attention is placed more on pushing for the legalization of prostitution as work. In fact, during the UN sessions on drafting the UN Protocol, the most contentious debate was among feminists about how to determine what "exploitation" and "consent" should be within the activity of prostitution. As reported by Doezema, one lobby group "framed prostitution as legitimate labor. The other group considered all prostitution to be a violation of her human rights" (2005: 62). Unfortunately, while feminists on the opposing side of the debate passionately argue either to advocate criminalization of the client or to legitimize the sex trade as a form of labor, all evidence shows that females continue to be victims trafficked across borders into the sex industry.

Still "stuck in a groove": Gender, politics and cross-border sex trafficking

The ILO estimates that 55 percent of forced labor victims are women and girls. In total, approximately 37 percent of all human trafficking is for sexual exploitation, of which 98 percent are women and girls. The ILO also estimated a higher percentage of sex trafficking victims in 2012 than in 2005 (USDOS 2012).[7] In this chapter, I argue that the UN anti-trafficking guidelines are centered on the "victims," who are either adult or child females who have been trafficked into the commercial sex industry. They remain voiceless and practically anonymous; they are hidden behind a highly clandestine industry; and they are controlled and used for the most part by profit-seeking pimps, criminal groups, brothel owners, leisure industry groups and military groups (Enloe 2000; Whitworth 2004).

In this study, I have shown that, driven by conflicting socio-economic interests of source (global North) and destination (global South), the countries of the global North and the global South follow different agendas and policies on international migration. The sex trafficking flow of women and girls is an unwanted and stigmatized byproduct, unleashed by forces of globalization (Kapur 2003; Sassen 2000; Samarasinghe 2003). Immigration policies of sending and receiving countries seem to either ignore or work around the issue of "victimhood" of the trafficked female into the commercial sex sector. In general, sending countries tend to ignore the issue because of their dependence on remittances based on their labor export policy. In contrast, in receiving countries, notwithstanding their anti-sex trafficking policies, protection of their borders against illegal migration seems to take precedence over any systematic efforts to identify "victims" of cross-border sex trafficking.

The prostitution/sex work debate superimposed on the already complex migration/trafficking nexus, creates yet another set of divisive issues that pose further challenges to formulating effective policy measures to eliminate female sex trafficking. Identifying the "victim" has led to a situation where groups agree that there are victims of sexual exploitation; however, they do not agree about whether or not "victims" are all females engaged in prostitution or whether their victimhood in sex trafficking is an essential component of labor exploitation. In this case, adult women should have the right to use the sale of sex as a form of labor. Thus, while the UN Protocol urges member nations to adopt "victim-centered" anti-sex trafficking initiatives, designing such a cohesive set of effective anti-sex trafficking policies is stalling in the face of the challenges encountered in identifying the female "victim" of sex trafficking.

Notes

1 The three "Ps" were originally created in 1999 by the US government under the Clinton administration.

2 See Jordan (2002).
3 International migration in this context includes both documented and undocumented migrants.
4 The four top remittance-receiving countries are: India, Mexico, the Philippines and China. In India, Mexico and China, male migrants outnumber female migrants (Ratha *et al.* 2012).
5 In 2004, an activist in an anti-trafficking NGO in Phnom Penh (in the course of a conversation on corruption of law enforcement) argued that police officers who are paid poor salaries are tempted to augment their income by receiving bribes from pimps and traffickers. She said, "When you pay peanuts, you get monkeys."
6 The U.S. TIPS Report identifies three tiers in its ranking of countries on anti-sex trafficking initiatives, Tier 1 is the best and Tier 3 is the worst in terms of the effectiveness of state policies in curbing this practice. Rankings are based on the assessment of the degree to which each country complies with the minimum standards identified in the US Trafficking in Persons Act of 2000 (USDOS 2012).
7 The ILO recognizes that human trafficking is defined by exploitation and not by movement. As quoted in the TIPs Report (USDOS 2012), the statistics are derived from ILO estimates as of June 1, 2012.

References

Agustin, L. M. (2007) *Sex at the Margins: Migration, Labor Markets and the Rescue Industry*, London: Zed Books.

Anderson, B. and Davidson J. (2004) *Trafficking—A Demand Led Problem?* Sweden: Save the Children.

Apap, J. (2003) *Protection Schemes for Victims of Trafficking in Selected EU Member Countries, Candidates and Third Countries*, Geneva: IOM.

Bales, K. and Lise, S. (2005) "Trafficking in Persons in the United States," document No. 211980, Washington, DC: National Institute of Justice.

Barry, K. (1995) *The Prostitution of Sexuality*, New York: New York University Press.

Beloso, B. M. (2012) "Sex, work, and the feminist erasure of class," *Signs* 38(1): 47–70.

Bernstein, E. (2010) "Militarized humanitarianism meets carceral feminism: the politics of sex rights, and freedom in contemporary anti-trafficking Campaigns," *Signs* 36(1): 45–71.

Cameron, S. (2008) "Trafficking of women for prostitution," in S. Cameron and E. Newman (eds) *Trafficking in Humans: Social, Cultural and Political Dimensions*, Tokyo: United Nations University Press, pp. 80–110.

Cheung, J. (2006) "The United States as global sheriff: using unilateral sanctions to combat human trafficking," *Michigan Journal of International Law* 27: 437–494.

Coalition Against Trafficking of Women (CATW) (1998) *Philosophy of the Coalition Against Trafficking of Women*, CATW: Manila.

Cornwall, A., Harrison, E. and Whitehead, A. (eds) (2007) *Feminisms in Development: Contestations and Challenges*, London and New York: Routledge.

Di Nicole, A. (ed.) (2004) "A study of monitoring the international trafficking of human beings for the purpose of sexual exploitation in the member eu states," Transcrime. Available online at: transcrime.cs.unitn.it/tc/423.php (accessed May 31, 2013).

Doezema, J. (2005) "Now you see her, now you don't: sex workers at the UN trafficking protocol negotiations," *Social and Legal Studies* 14(1): 61–89.

Doezema, J. (2010) *Sex Slaves and Discourse Masters: The Construction of Trafficking*, London and New York: Zed Books.

Enloe, C. (2000) *Maneuvers: The International Politics of Militarizing Women's Lives*, Berkeley, CA: University of California Press.

Global Alliance Against Trafficking of Women (GAATW) (1994) Available online at: www.gaatw.org (accessed May 28, 2013).

Hesford, W. and Kozol, W. (eds) (2005) *Just Advocacy?: Women's Human Rights, Transnational Feminisms, and the Politics of Representation*, New Brunswick, NJ: Rutgers University Press.

Hopkins, R. and Nijboer, J. (2004) "Trafficking in human beings and human rights: research policy and practice in the Dutch approach," *Human Rights Law Review* (Special Issue Spring 2004): 75–89.

Hua, J. (2011) *Trafficking Women's Human Rights*, Minneapolis, MA: University of Minnesota Press.

Human Rights Watch (2000) *Owed Justice: Thai Women Trafficked Into Debt Bondage in Japan*, Brussels: Human Rights Watch.

International Labor Organization (ILO) (2005) *Human Trafficking for Sexual Exploitation in Japan*, Tokyo: ILO.

Jeffreys, S. (2009) *The Industrial Vagina*, Oxford and New York: Routledge.

Jordan, A. (2002) *The Annotated Guide to the Complete UN Trafficking Protocol*, Washington, DC: International Human Rights Law Group.

Kapur, R. (2003) " 'The other side of globalization': the legal regulation of cross-border movement," *Canadian Women's Studies* 22(3/4): 6–15.

Kapur, R. (2008) "Migrant women and the legal politics of anti-trafficking initiatives," in S. Cameron and E. Newman (eds) *Trafficking in Human Beings: Social, Cultural and Political Dimensions*, Tokyo: United Nations University Press, pp. 111–25.

Kempadoo, K. (2005) "Introduction: from moral panic to global justice: changing perspectives on trafficking," in K. Kempadoo with Jyoti Sanghera and Bandana Pattanaik (eds) *Trafficking and Prostitution Reconsidered: New Perspectives on Migration, Sex Work and Human Rights*, Boulder, CO: Paradigm Publishers, pp. viii–xxxiv.

Kondo, A. (2011) "Japanese experience and response in combating trafficking," in S. Okubo and L. Shelley (eds) *Human Security, Transnational Crime and Human Trafficking: Asia and Western Perspectives*, New York: Routledge, pp. 216–32.

Martilla, A. (2003) "Consuming sex—Finnish male clients and Russian Baltic prostitution," paper presented at Gender and Power in New Europe, the 5th European Feminist Research Conference, Lund University, Sweden, August 20–24.

Outschoorn, J. (2004) "Voluntary and forced prostitution: the 'realistic approach' of the Netherlands," in J. Outschoorn (ed.) *The Politics of Prostitution: Women's Movements, Democratic States and the Globalisation of Sex Commerce*, Cambridge: Cambridge University Press, pp. 165–204.

Pearson, E. (2002) Human Traffic, Human Rights: Redefining Victim Protection. London: Anti-Slavery International.

Ratha, D., Mohopatra S. and Silwal, A. (2012) *Migration and Remittances: Fact Book 2011*, Migration and Remittances Unit, Washington, DC: World Bank. Available online at: www.Worldbank.org/prospects/migrationandremittances (accessed May 28, 2013).

Raymond, J. (2004) "Prostitution on demand: legalizing the buyers as sexual consumers," *Violence Against Women* 10(10): 1156–86.

Samarasinghe, V. (2003) "Confronting globalization in anti-trafficking strategies in Asia," *The Brown Journal of World Affairs* 10(1): 91–104.

Samarasinghe, V. (2008) *Female Sex Trafficking in Asia: Resilience of Patriarchy in a Changing World*, New York: Routledge.

Sanders, T. and Campbell, R. (2008) "Why men hate paying for sex: exploring the shift to tackling demand in the U.K.," in E. V. Munroe and M. D. Giusta (eds) *Demanding Sex: Critical Reflections on the Regulation of Prostitution*, New York: Ashgate.

Sassen, S. (2000) "Women's burden: counter geographies of globalization and the feminization of survival," *Journal of International Affairs* 53(2): 504–23.

Scarpa, S. (2008) *Trafficking in Human Beings: Modern Slavery*, Oxford: Oxford University Press.

Shelley, L. (2010) *Human Trafficking: A Global Perspective*, Cambridge: Cambridge University Press.

Svanstrom, Y. (2004) "Criminalizing the John—a Swedish gender model," in J. Outschoorn (ed.) *The Politics of Prostitution: Women's Movements, Democratic States and the Globalization of Sex Commerce*, Cambridge: Cambridge University Press, pp. 225–44.

UNODC (United Nations Office on Drugs and Crime) (2003) "Coalition against trafficking in human beings in the Philippines: research and action," Final Report, Vienna: Global Program Against Trafficking in Human Beings, Anti-Human Trafficking Unit.

United States Department of State (USDOS) (2008) *Trafficking in Persons Reports*, Washington, DC: USDOS.

United States Department of State (USDOS) (2012) *Trafficking in Persons Reports*, Washington, DC: USDOS.

Whitworth, S. (2004) *Men, Militarism and UN Peacekeeping: A Gendered Analysis*, Boulder, CO: Lynne Reinner.

Wijers, M. and Lap-Chew, L. (1997) *Trafficking in Women, Forced Labor and Slavery-Like Practices in Marriage, Domestic Labor and Prostitution*, Netherlands: Foundation Against Trafficking in Women (STV).

Zhang, S. X. (2007) *Smuggling and Trafficking in Human Beings: All Roads Lead to America*, London: Praeger.

Part II

Gendering the field

Participatory feminist research

5 Crossing boundaries

Transnational feminist methodologies in the global North and South

Ann M. Oberhauser

> We drove to the house in a remote, rural area to meet women who were members of a state-wide knitting cooperative. As we approached, I saw a large garden and clothes hanging on the wash line in the back yard. Several women came out to greet us, curious about our visit and the purpose of our research project. We spent the next few hours talking to these women about their cooperative and how they integrate their economic activities with domestic responsibilities.
>
> (Field notes—West Virginia, US 1997)

> The thatched huts and brick buildings were surrounded by fields of maize and grazing areas for local cattle. Members of the Pfanani Cooperative were in the main building, weaving cloth at their looms and putting handbags, wall hangings and sweaters into boxes for transport to the local market. Several women had babies on their backs or watched toddlers play in the yard. We toured the production and storage areas before sitting down to discuss their economic activities.
>
> (Field notes—Limpopo Province, South Africa 2001)

Introduction

These notes, compiled from research conducted in two seemingly disparate study sites, reflect remarkably similar observations and experiences in the field. The rural settings and evidence of subsistence livelihoods illustrate intersecting aspects of women's economic strategies in a transnational context. The collective economic strategies of participants in this research form the basis for my analysis of gendered livelihoods in the two rural areas of Limpopo Province (South Africa) and the Appalachian Region (US). The methodological focus of this chapter draws from transnational feminism in order to highlight intercultural relations between researchers and participants across transnational boundaries (Aitken 2010; Skelton 2001; Nagar 2002). Transnational feminism engages with themes of difference and praxis in global fieldwork by emphasizing similar political and economic practices among people from diverse backgrounds in multiple global locations (Pratt 2012; Sangtin and Nagar 2006; Peake and Trotz 1999). Through these fluid and contested boundaries, people, traditions and ideas become mobile and interconnected within the context of globalization. Furthermore, transnational feminism challenges conventional categories and boundaries as being too restrictive and static in their representation of global structures and provides an alternative framework that focuses on the intersections of social and spatial processes in intercultural research (Khagram and Levitt 2008).[1]

The framework adopted in this analysis underscores the socio-economic dimensions of cross-boundary and intercultural connections that shape many aspects of the research process. The discussion draws from Cindi Katz's conceptualization of topographical contour lines in research that enables:

> the examination of relationships across spaces and between places. The material social practices associated with globalization work in interconnection...but they work iteratively as well: the effects of capitalism's globalizing imperative are experienced commonly across very different locales, and understanding these connections is crucial if they are to be challenged effectively.
>
> (Katz 2001: 9)

Counter-topographies expand this concept of topographical contour lines to reflect the situated responses and formation of political-economic alliances and strategies that contest conventional economic globalization.

Transnational feminism offers a critical methodological framework to deconstruct and challenge hierarchies between the global and the local, the separation of public and private spaces and other dualisms that are constructed across multiple scales. This framework also examines difference and power relations among researchers and participants in the research process and offers alternative ways of thinking about how these power relations affect knowledge production and distribution of resources (Kobayashi and Peake 1994; Mohammad 2001). This research draws from this approach to examine how participants are empowered through access to resources and support for income-generating activities.

The empirical component of this study focuses on women's livelihood strategies in two regions: South Africa and the US. The intersecting analysis of these economic strategies in both South Africa and the United States informs this research in ways that highlight their dynamic and contingent nature. Specifically, the home-based aspect of these livelihoods, and their participation in cooperative economic strategies, draw parallels that are locally situated yet transnationally connected to larger neoliberal global processes. In addition, the socio-economic processes and geographical sites associated with these livelihood strategies are examined in the context of power relations in the field. These strategies provide counter-topographies of struggles and alliances that shape the research process.

This chapter is organized into five sections. The second section examines transnational feminist methodology, with a focus on intercultural research and praxis, which take place in geographically separate and culturally diverse locations. The third section provides the empirical context of this research on gendered livelihoods in Appalachia and Limpopo. The socio-spatial relationships within and across the two case study regions underscore the importance of a transnational feminist analysis in exposing often contested economic strategies. The discussion in the fourth section highlights the role of these social and spatial differences and the transformation that occurs among both researcher and participants through intercultural fieldwork. The conclusion sheds light on the relevance of this analysis to transnational feminism and specifically how it informs and shapes methodological approaches in both feminism and critical geography.

Transnational feminism as methodology and fieldwork

Transnational feminism is useful in navigating multiple scales and challenging conventional boundaries that situate structures and processes within local, regional, global and nation-state systems (Swarr and Nagar 2010; Alexander and Mohanty 2010; Pratt and Yeoh 2003). This research approach engages with the intersection of alternative spaces and boundaries, such as those that occur in and among communities, households and individuals. Given the often dynamic and contested nature of these cross-cultural sites and participants, transnational feminism deconstructs dichotomies such as insider/outsider, theory/method and researcher/subject in efforts to reframe power relations that stem from the interaction between levels of social and economic experience (Khagram and Levitt 2008).

My research in lesser-developed regions of the global North and South adopts a transnational, postcolonial framework, where power relations are examined both within capitalist forces of domination and exploitation, and among participants. As a white female researcher, I embody some of the power relations that affect my interaction with rural women in parts of Appalachia, where I live and conduct research, as well as in Limpopo Province, where I have conducted research on numerous occasions during the past twenty years (Figure 5.1). My transnational engagement with the research process disrupts the dominant discourses of uneven development and inequality in order to contextualize these power relations. Likewise, dominant capitalist forces assign certain access to, and control over, resources among the participants that affect the intercultural and dynamic nature of these power relations. Within this global framework, one can trace the formation of political and economic strategies that provide counter-topographies for resisting oppressive and marginalizing forces of globalization.

Transnationalism has multiple applications in feminist work that include transgressing conventional interpretations of difference linked to unequal power relations (Nagar and Swarr 2010; Alexander and Mohanty 2010; Peake and de Souza 2010). Similarly, Young's (1990) "politics of difference" goes beyond the reification of difference in categories such as race, class, gender and sexuality. In an effort to broaden and deepen our understanding of difference, Young identifies several modalities of power involving issues of justice that include exploitation, marginalization, powerlessness, cultural imperialism and violence. According to Aitken, "by adding the complexities of multiple axes of oppression without diluting the force of oppressive practices, she provides a methodology for highlighting the myriad ways difference is articulated as a form of disenfranchisement"

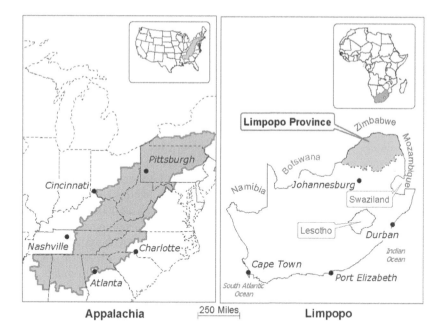

Figure 5.1 Appalachia, United States and Limpopo Province, South Africa. (Map by George Roedl, 2013.)

(2010: 59). Young thus opens up methodological possibilities that are relevant to intercultural research. Additionally, by incorporating activism and praxis, transnational feminism reveals often ignored spaces of power and resistance (such as the body, household and community), which contribute to the methodological richness of, and possibilities in, comparing and engaging difference.

This chapter draws from transnational feminist methodology to further advance our understanding and application of intercultural and transformative research. The analysis addresses issues of difference and collaboration in ways that challenge conventional thinking about socio-spatial aspects of fieldwork in cross-cultural contexts. Specifically, this methodological approach focuses on negotiating intercultural dynamics, uncovering power relations in the global North and South and incorporating praxis into the research process. The next section examines income-generating strategies among economically marginalized women in the geographically separate and culturally diverse research sites of Appalachia (US) and Limpopo Province (South Africa).

Economic strategies in Appalachia and South Africa

This analysis is based on research conducted in central Appalachia and northern South Africa, where gendered livelihood strategies occur within the context of neoliberal economic restructuring (Figure 5.1). The fieldwork is situated in both physical and social spaces that are shaped by and affect both participants and researchers.

Transnational feminist themes reveal complex aspects of the field, social difference and the role of shifting economic forces in these two rural, lesser-developed regions that have been shaped by neoliberal capitalism. The political economy of these areas is embedded in a history of natural resource extraction, exploitation of labor and people's marginalization from the land (Hart 2002; Billings and Blee 2000). This discussion examines the spatial and gendered aspects of these economic geographies before focusing on topographies (or connections) and relationships between participants and researchers across the study sites.

In the Appalachian Region, the extraction of coal, timber and natural gas has developed in local economies dominated by large-scale and often absentee landowners, corporations and capitalist market forces (Billings and Blee 2000). Indeed, Appalachia has been described as a third world region in a first world setting due to its dependence on external capital and exploitative labor relations (Lewis and Billings 1997). Shifts in market forces and means of production in the past thirty years have led to the growth of service sector jobs in many areas of the region alongside the decline of primary and secondary sector jobs (Latimer and Oberhauser 2004). West Virginia, the only state lying completely within the political boundaries defined by the Appalachian Regional Commission, exemplifies many of these regional economic trends. For instance, in this state, coal-mining jobs have decreased by the thousands in recent decades, due to the mechanization of coal mining and mountaintop removal, the growth of energy sources like natural gas and government regulations of coal-fired power plants (Woods 2011).

Gender plays a significant role in this economic restructuring, especially in terms of employment trends and income disparities. In West Virginia, women's labor force participation has increased during recent decades with nearly 60 percent of women working in 2010, compared to 34 percent in 1950 and 43 percent in 1970 (US Department of Commerce 2010). The increase in the number of women who work outside of the home reflects in part the growing number of employment opportunities, particularly in the service sector, and the need for additional household income (Latimer and Oberhauser 2004). However, growth in female employment has not improved rates of pay for women: West Virginian women earn 69 percent of men's earnings, compared with 79 percent in the US as a whole (US Department of Commerce 2010).

South Africa's Limpopo Province has also undergone significant economic shifts in recent decades, with post-apartheid reforms and neoliberal globalization. Uneven development and social inequality in South Africa result from colonial structures and white minority rule under apartheid during the last several centuries. In many rural areas, blacks were forced to live on marginal land in the former Bantustans, with poor access to resources and few opportunities for viable economic activities (Lipton *et al.* 1996). Women were particularly disadvantaged under the apartheid regime, with limited rights to land ownership and marginalization from other resources and capital assets (Meer 1997; Francis 2000).

Government efforts to expand economic opportunities, and to overcome social inequalities that formed under colonialism and apartheid, continue to confront challenges posed by neocolonial and neoliberal capitalist forces (Magubane 2004). Rural black Africans in particular face serious economic hardships as the government implements a market-based program of land reform to maintain agricultural competitiveness and to ensure adequate production in a region with growing food insecurity (Walker 2002; Tsheola 2012). Limpopo is one of the poorest provinces in South Africa, with over 80 percent of the population living in rural areas and an unemployment rate of 32.4 percent in 2007 (TCOE 2012). Women remain especially vulnerable to these economic hardships for reasons that include cultural norms and values about gender and work. As Walker explains:

> both custom and law have underpinned rural women's economic marginalization. Although no longer constitutionally defined as perpetual minors in the eyes of the law, women continue to be treated as subordinate to men, and many parties defend this subordination in the name of tradition and African culture.
>
> (Walker 2002: 73)

These prevailing attitudes have contributed to a situation where poverty rates are highest among women, particularly in rural areas that have experienced the outmigration of male laborers for decades (Francis 2000; Tsheola 2012). According to a recent study of gender and economic status in South Africa, 60.4 percent of women in rural settlements in Limpopo Province live in poverty (Kongolo 2009).

Attention to uneven development and social inequality in both case study regions of the US and South Africa reinforces the important role of alternative economic strategies in household livelihoods. These strategies include barter exchange, subsistence production, micro-enterprise development and other means of generating income or subsistence for households (Oberhauser 2002, 2005; Tsheola 2012). Many of these alternative economies occur in the context of economic restructuring, especially the shift to service sector jobs as employment in mining and manufacturing decline in many parts of Appalachia and northern South Africa. Many Africanists also document the role of neoliberal capitalism and market forces within the post-apartheid government as contributing to these economic shifts (Ferguson 2006; Magubane 2004).

Household dynamics and gender roles are important aspects of these alternative economic strategies in both the Appalachian Region and Limpopo Province (Oberhauser 2005). In particular, women's marginalization from formal labor and engagement in informal and home-based economic livelihoods are shaped by traditional roles and divisions of labor that confine women to domestic spaces and reproductive labor (Tsheola 2012; Francis 2000). Moreover, many individuals and households face limited economic opportunities and lack mobility in these rural areas. These barriers are due to not only the rural terrain, but also the lack of investment in basic infrastructure such as roads and means of transportation.

These conditions prevent people from accessing resources as well as markets and material inputs to their income-generating activities.

In this analysis, geographical context is related to the socially situated positions of both the researcher and the participants who often have diverse socio-economic, cultural and other backgrounds. My ability to work within and across multiple geographical contexts with various groups of people, despite their differences, leads to rich understandings of places and social processes. Feminist geographers have conducted extensive intercultural (or in some cases cross-cultural) research in multiple geographic locations on gendered aspects of political economic trends. For example, in her analysis of topographies of globalization in Howa Village (the Sudan) and Harlem (New York), Katz (2004) demonstrates how methodological intersections of research create opportunities that go beyond the comparison of conventional indices and socio-spatial patterns. Hart also examines the constitutive forces of political economy that act across spatial scales as the "accelerating global flows and connections" and the "specific articulations of difference that actively shape diverse trajectories of socio-spatial change" (2002: 37). Likewise, intersections of individual, household and community scales in the development of social networks affect the practice of alternative economic strategies in both Appalachia and Limpopo Province. Specifically, the cooperatives and producer networks in this project have formed around collective economic activities such as sewing, agriculture, pottery and other ventures that bring women together in these communities (Oberhauser 2005; Oberhauser and Pratt 2004).

In sum, the context for this research involves two case study areas with similar socio-spatial processes that contribute to their economic marginalization under hegemonic capitalist forces. Alternative livelihoods in these areas are partly a response to neoliberal economic reforms that continue to leave people with limited opportunities for viable economic strategies within the formal workforce (Tsheola 2012; Billings and Blee 2000; Oberhauser 2005). Transnational feminism thus informs our understanding of alternative economic strategies and, particularly, the importance of intercultural analysis and praxis. The topographical contour lines outlined in this research reflect important relationships across spaces and between places that are discussed in the methodological section below.

Practicing transnational feminist methodology

Researching gender and economic livelihoods in diverse geographical contexts highlights the crucial role of methodological approaches that incorporate intercultural relations, reveal important axes of difference and promote praxis and activism. These themes are discussed here in reference to socio-spatial relationships and processes across the global South and North. As described in the previous section, the economic marginalization of women in Appalachia and Limpopo Province reflects parallel processes of inequality and uneven development in the context of broader neoliberal capitalism. Moreover, transgressive economic activities or alternative livelihoods have developed outside the formal sector through cooperatives, informal work and other alternative economic strategies (Gibson-Graham 2006). These strategies are part of the counter-topographies or contested sites of resistance (Katz 2001; 2004) that inform this methodological approach and, consequently, of my intercultural interaction with participants in the field.

Transnational feminist research crosses boundaries and navigates intersecting scales to better understand intercultural practices and processes (Peake and de Souza 2010). The two study sites described above are seemingly disparate, yet they reveal intersecting and similar strategies among rural women. This intercultural research challenges the tendency to rely solely on conventional indices and socio-economic measures (such as per capita income and unemployment) and instead addresses the experiences, skills and knowledge that women have gained in these geographical contexts. The participants, settings and economic strategies observed in both regions are part of this cross-boundary research that creates shared experiences among participants and, consequently, lays the foundation for countering the ways that "globalized capitalism exacerbates and builds upon gendered, racialized, nationalist and class axes of oppression and inequality in different historical geographies" (Katz 2001: 10). In an attempt to connect these activities across space and scale, this methodological analysis focuses on everyday practices in the household, between individuals and within the community, as a means of understanding and analyzing "the field." These geographically disparate yet similar contexts and analyses, lead to a more thorough understanding of how households cope in the face of economic uncertainty.

In this research, households are important sites of fieldwork and engagement with female participants who conduct income-generating activities in their homes. Within this household space, gender relations are critical to economic strategies and their intersection with reproductive labor. At the community scale, several of the producer groups in South Africa work in areas within and beyond the household that include community gardens, fire circles, grazing areas and other communal spaces. For example, the Tshandama Community Peanut Butter and Jam Project roasted peanuts gathered from a local field in fire circles outside their homes. This and other economic activities took place in communities and households while caring for children and conducting other domestic responsibilities (Figure 5.2).

Similar situations are evident in the Appalachian fieldwork, where economic activities and domestic responsibilities overlap in household spaces. One woman described the shared space that she used for her knitting in her trailer home, where she has her sewing machine in the bedroom (Figures 5.3[a] and [b]):

My husband accuses me of taking over the bedroom…because the way the trailer is designed, food odors are a hazard to our yarn. So our bedroom happens to be the furthest away from the kitchen area. We had talked about the living room…where I could be with the kids while they are home. But the food odors became a real problem with that.

Figure 5.2 Members of the Tshandama Community Project. (Photo by Ann M. Oberhauser, 2002.)

These connections and shared experiences shed light on processes and, in turn, strategies of resistance and alternative economic praxis. With reference to the challenges of combining domestic labor with economic activities in her home, another Appalachian woman stated:

> you work around it. 'Cause like now, the children are out of school for the summertime. I was thinking of knitting at night time or in the early morning or before they're up.... (It's important) to work around the kids because the kids are demanding.

In these examples, home-based work that entails growing vegetables, knitting clothes and processing food occurs alongside domestic labor and within household spaces.

This analysis illustrates how intercultural, transnational research enhances our understanding of the socio-economic strategies and dynamics that comprise the field. My observations of shared economic strategies and household responsibilities among the participants raise methodological issues about where and how one conducts fieldwork. During this fieldwork, I entered women's households or their communal space and conducted interviews as they engaged in paid labor and conducted household chores such as childcare, cooking or cleaning. As highlighted in the field notes at the beginning of this paper, my approach to the field was embedded in various social and spatial dynamics that required careful negotiation across different scales and relationships (Aitken 2010; Skelton 2001). The research process also included instances where difference was manifest in my interactions with the participants. These differences stem from our social and economic backgrounds that reflect diverse configurations of power frequently evident in intercultural research (Skelton 2001). As Young (1990) states, difference is

Figure 5.3 (a) Knitter's work space; (b) Knitter's home in rural West Virginia. (Photos by Ann M. Oberhauser, 1997.)

defined through processes that entail exploitation, marginalization, powerlessness and cultural imperialism. In many cases, these differences are unavoidable but, nonetheless, they affect interactions among participants and overall dynamics of engaging in this research.

The intercultural research outlined in this discussion involves multiple axes of difference across diverse social and economic geographies. For example, the differences between me as a white female academic and participants in rural Appalachia and Limpopo Province were evident and contributed to the complicated and sometimes contested dynamics within this intercultural research. Many women who participated in the study did not have a college education or a high school degree. These fieldwork experiences parallel Aitken's (2010) discussion of "encountering" different peoples and, in many cases, working within the socially constructed axes of difference. Encounters such as these are built on historically-grounded systems of colonialism, racism, capitalism and other types of exploitation and uneven development.

The themes in this analysis reflect participants' perceptions of difference or power dynamics in fieldwork stemming from unequal access to material goods, financial resources and mobility. In turn, these perceptions raise expectations among the participants about how research can help them expand their access to resources and other opportunities. For example, I was approached by participants in Appalachia and South Africa about financial support for agricultural gardens, knitting cooperatives and livestock farms. Some participants also asked about my ability to provide scholarships and grants for young people in their families to enable them to enroll in the university. I responded to these inquiries by (re)stating the purpose of the research and my positions as both an academic researcher and a collaborator with community organizations. In some cases, collaboration with community groups and university staff established connections for the participants to gain access to potential resources and expertise. Nonetheless, some of the research was perceived as being embedded in neocolonial relations that reflect disparities and unequal access to resources and, ultimately, the production of knowledge (Alexander and Mohanty 2010).

Finally, transnational feminism includes praxis or advocacy as a way to counter some of these inequalities. In some cases, this aspect of the research process empowers those participants involved in collaboration with academics and activists. The transformation and empowerment of participants in the research project entails engagement with women and community groups as a means of drawing out and learning from their experience. In light of these obligations and practices in transnational feminist methodology, I worked with cooperatives to share my findings and to generate results for their use. This process also informed me about the priorities of the participants, their interaction with outside groups and the policy issues that pertain to their economic activities. For example, I participated in meetings organized by the Appalachian knitting cooperative, which trained members in sewing techniques as well as in the financial management of their operations (Figure 5.4). Research findings and data on membership and activities have also been used to support grant writing activities and fundraising efforts of these groups. Likewise, my involvement with the South African cooperatives and small businesses established linkages with government agencies, such as the Rural Development Bank, which provided training and support for business development (Oberhauser and Pratt 2004). The groups also strengthened their connection to university researchers and government groups through their involvement in this research project.

Many feminist geographers have engaged with this type of praxis and collaborative research in intercultural projects. Geraldine Pratt's project with organizations representing migrant labor in Canada includes advocacy and intensive participatory research (Pratt 2012). Researchers in this project work closely with government officials in lobbying to improve the situation for migrant workers in Canada. Likewise, the Sangtin and Nagar (2006) developed an innovative project on transnational feminism that spans two continents and focuses on women's

Figure 5.4 Training session with knitters. (Photo by Ann M. Oberhauser, 2000.)

empowerment through a grass-roots NGO in India. Their collaborative methodology provides a unique approach to autobiographical writing that informs the authors' critiques of societal structures at all scales of analysis.

My research involves approaches similar to those outlined above through its collaboration with community groups, non-profit organizations, agricultural extension agents and academic researchers. In some cases, these groups provided access into the communities and allowed me to establish contacts with other groups. For example, the agricultural extension office in Limpopo Province includes a cooperative that raised pigs and served as a field site for this research (Oberhauser and Pratt 2004). The cooperative was receptive to feedback and input about their operations and economic strategies among members of the piggery. The relationship with groups such as the piggery plays an important role in our understanding of and efforts to critique gendered livelihoods in both Appalachia and Limpopo Province. In many respects, collaboration legitimized my presence in these communities where entry was perceived with curiosity and, in some cases, suspicion. I worked with the participants and these organizations to challenge discriminatory practices that denied rural women (and their households) access to resources and economic opportunities. Thus, the community groups and participants are positioned as not just "key informants," but also collaborators in the research process and, ultimately, in the goals and outcomes of the research.

Overall, the transnational component of this research involves critical analysis of the socio-economic context of the field(s), the interaction with participants and the negotiation of outcomes for transformation and change. Both of the research locations in the US and South Africa represent important sites for intercultural relationships and counter-topographies across similar spatial and social processes.

Analysis of these contexts using transnational feminist methodology thus reveals shared experiences that transcend borders and traverse scales.

Conclusion: Engaging transnational feminist research

Transnational feminism is part of a broader effort to develop relevant and effective intercultural research that promotes empowerment and progressive change among women and other marginalized groups. This approach also addresses growing socio-economic and spatial inequalities in both the global South and North. The methodological framework that is embedded within transnational feminism is geared to issues of difference, power relations and praxis that affect fieldwork and the research process as a whole. Moreover, transnational methods challenge conventional approaches that confine people and places to specific categories and bounded territories (such as the region or nation-state) and focus instead on transnational and cross-border flows, linkages and identities (Khagram and Levitt 2008). As a whole, this framework encourages researchers to reinvent themselves (and others) in order to analyze the progressive potential of unbounded space and liberating social processes.

The methodology outlined in this chapter draws from several themes that constitute transnational feminist research. The discussion aims to promote critical analysis of the social processes and power relations that shape the field and, in particular, intercultural interactions among researchers and participants. In addition, the transformative component of this framework adopts themes that align with feminist activism and advocacy in the research process (Alexander and Mohanty 2010). Transnational feminism often involves collaborative projects that actively engage both researchers and participants across political, cultural and economic boundaries. These collaborations, in turn, shift power relations and specific axes of difference that are embedded in conventional approaches toward more mutually beneficial outcomes (Swarr and Nagar 2010).

This research on gendered livelihoods involves collaboration with cooperatives and other community organizations in the Appalachian Region and Limpopo Province that work to provide opportunities and resources for economically vulnerable women. The study areas reflect sites of parallel experience, yet they contain place-specific challenges to efforts to empower individuals, households and communities. Moreover, the intersection of gender with other social identities and, more importantly, differences in the form of exploitation, marginalization, powerlessness and cultural imperialism (Young 1990) creates specific challenges in these material settings. Thus transnationalism does not replicate similar processes in particular places but examines (through counter-topographies) the layers of processes and their intersections with social practices at other scales (Katz 2001).

Intercultural analysis enriches this research project by highlighting the intersections of socio-spatial processes in the geographic contexts of the Appalachian Region and Limpopo Province. Unequal gender roles and divisions of labor are evident in both study sites, albeit with specific nuanced differences and sometimes contrasting outcomes. For example, cooperatives and collective economic strategies in both regions have very different outcomes depending on labor practices, economic policies and support from government and non-governmental organizations. In addition, the remote, rural and often impoverished geographic context shapes this analysis of gendered livelihoods in particular and specific ways. In some instances, these geographic conditions alienate and marginalize women while other women overcome these challenges and build viable businesses with community-based support. The integrated analysis of counter-topographical feminist fieldwork provides a unique lens to view and understand these processes.

Finally, this discussion highlights the role of praxis and transformation in transnational feminist research. Through empowering methodological practices, change can occur that will further build communities, strengthen individuals and develop equality in households. These practices entail working with a broad spectrum of partners and collaborators such as community-based groups and local resources. Transnational feminism traces and disrupts historically embedded axes of difference and inequality that often shape intercultural fieldwork. The scenes described in the opening paragraphs of this chapter illustrate the parallel and profound encounters that researchers face in transnational fieldwork that crosses national boundaries and is situated in seemingly diverse and geographically disparate communities. Analyses of gendered livelihoods in rural areas of the US and South Africa thus go beyond comparative studies of economic activities to critically engage with intercultural socio-spatial processes in globally-integrated research settings. This approach provides opportunities for feminist oppositional politics to promote progressive change at multiple scales.

Note

1   I use the term "intercultural" in this discussion to emphasize collaborative and relational connections in the research process. In contrast, the concept "cross-cultural" suggests ways to understand or represent another culture in a comparative fashion that often set up dualisms or hierarchical relations. In this research, intercultural pulls together diverse cultural forms among the participants towards the creation of something new.

References

Aitken, S. (2010) " 'Throwtogetherness': encounters with difference and diversity," in D. DeLyser, S. Herbert, S. Aitken, M. Crang, and L. McDowell (eds) *The SAGE Handbook of Qualitative Geography*, London: SAGE, pp. 46–68.

Alexander, M. J. and Mohanty, C. T. (2010) "Cartographies of knowledge and power: Transnational feminism as radical praxis," in A. L. Swarr and R. Nagar (eds) *Critical Transnational Feminist Praxis*, Albany, NY: SUNY Press, pp. 23–45.

Billings, D. and Blee, K. M. (2000) *The Road to Poverty: The Making of Wealth and Hardship in Appalachia*, Cambridge: Cambridge University Press.

Ferguson, J. (2006) *Global Shadows: Africa in the Neoliberal World Order*, Durham, NC: Duke University Press.

Francis, E. (2000) *Making a Living: Changing Livelihoods in Rural Africa*, London: Routledge.

Gibson-Graham, J. K. (2006) *The End of Capitalism (As We Knew It): A Feminist Critique of Political Economy*, Minneapolis, MN: University of Minnesota Press.

Hart, G. (2002) *Disabling Globalization: Places of Power in Post-Apartheid South Africa*, Pietermaritzburg, SA: University of Natal Press.

Katz, C. (2001) "On the grounds of globalization: a topography for feminist political engagement," *Signs* 26(4): 1213–34.

Katz, C. (2004) *Growing Up Global: Economic Restructuring and Children's Everyday Lives*, Minneapolis, MN: University of Minnesota Press.

Khagram, S. and Levitt, P. (2008) "Constructing transnational studies," in S. Khagram and P. Levitt (eds) *The Transnational Studies Reader: Intersections and Innovations*, New York: Routledge, pp. 1–18.

Kobayashi, A. and Peake, L. (1994) "Unnatural discourse: race and gender in geography," *Gender, Place and Culture* 1: 225–43.

Kongolo, M. (2009) "Women in poverty: experience from Limpopo Province, South Africa," *African Research Review* 3(1): 246–58.

Latimer, M. and Oberhauser, A. M. (2004) "Exploring gender and economic development in Appalachia," *Journal of Appalachian Studies* 10(3): 269–91.

Lewis, R. L. and Billings, D. B. (1997) "Appalachian culture and economic development: a retrospective view on the theory and literature," *Journal of Appalachian Studies* 3(1): 3–42.

Lipton, M., Ellis, F. and Lipton, M. (eds) (1996) *Land, Labour and Livelihoods in Rural South Africa: Vol 2, KwaZulu-Natal and Northern Province*, Durban: Indicator Press.

Magubane, A. (2004) "The revolution betrayed?: globalization, neoliberalism, and the post-Apartheid state," *The South Atlantic Quarterly* 1034: 657–71.

Meer, S. (ed.) (1997) *Women, Land and Authority: Perspectives from South Africa*, Oxford: Oxfam.

Mohammad, R. (2001) " 'Insiders' and/or 'outsiders': positionality, theory and praxis," in M. Limb and C. Dwyer (eds) *Qualitative Methodologies for Geographers: Issues and Debates*, London: Oxford University Press, pp. 101–17.

Nagar, R. (2002) "Footloose researcher, traveling theories, and politics of transformative feminist praxis," *Gender, Place, and Culture* 9(2): 179–86.

Nagar, R. and Swarr, A. L. (2010) "Theorizing transnational feminist praxis," in A. L. Swarr and R. Nagar (eds) *Critical Transnational Feminist Praxis*, Albany, NY: SUNY Press, pp. 1–20.

Oberhauser, A. M. (2002) "Relocating gender and rural economic strategies," *Environment and Planning A* 34: 1221–37.

Oberhauser, A. M. (2005) "Scaling gender and diverse economies: perspectives from Appalachia and South Africa," *Antipode* 37(5): 863–74.

Oberhauser, A. M. and Pratt, A. (2004) "Women's collective economic strategies and transformation in rural South Africa," *Gender, Place and Culture* 11(2): 209–28.

Peake, L. and de Souza, K. (2010) "Feminist academic and activist praxis in service of the transnational," in A. L. Swarr and R. Nagar (eds) *Critical Transnational Feminist Praxis*, Binghampton, NY: SUNY Press, pp. 105–23.

Peake, L. and Trotz, D. A. (1999) *Gender, Place and Ethnicity: Women and Identities in Guyana*, London: Routledge.

Pratt, G. (2012) *Families Apart: Migrant Mothers and the Conflicts of Labor and Love*, Minneapolis, MN: University of Minnesota Press.

Pratt, G. and Yeoh, B. (2003) "Transnational (Counter) Topographies," *Gender, Place and Culture* 10(2): 159–66.

Sangtin Writers and Nagar, R. (2006) *Playing with Fire: Feminist Thought and Activism Through Seven Lives in India*, Minneapolis, MN: University of Minnesota Press.

Skelton, T. (2001) "Cross-cultural research: issues of power, positionality and 'race'," in M. Limb and C. Dwyer (eds) *Qualitative Methodologies for Geographers: Issues and Debates*, London: Oxford University Press, pp. 87–100.

Swarr, A. L. and Nagar, R. (eds) (2010) *Critical Transnational Feminist Praxis*, Albany, NY: SUNY Press.

Trust for Community Outreach and Education (TCOE) (2012) "Limpopo Province." Available online at: http://tcoe.org.za/areas-of-operation/42-limpopo-province/71-limpopo-province.html (accessed May 21, 2013).

Tsheola, J. (2012) "Rural women's survivalist livelihoods and state interventions in Ga-Ramogale village, Limpopo Province," *African Development Review* 24(3): 221–32.

US Department of Commerce (2010) "Bureau of the Census, American Community Survey," Calculations by the Institute for Women's Policy Research based on S. Ruggles, J. T. Alexander, K. Genadek, R. Goeken, M. B. Schroeder, and M. Sobek, "Integrated Public Use Microdata Series: Version 5.0" (Machine-readable database), Minneapolis, MN: University of Minnesota, 2010. Available online at: http://usa.ipums.org/usa/ (accessed May 21, 2013).

Walker, C. (2002) "Land reform and the empowerment of rural women in postapartheid South Africa," in S. Razavi (ed.) *Shifting Burdens: Gender and Agrarian Change Under Neoliberalism*, Bloomfield, CT: Kumarian Press, pp. 67–92.

Woods, B. (2011) "Mountaintop removal and job creation: exploring the relationship using spatial regression," *Annals of the Association of American Geographers* 101(4): 806–15.

Young, I. M. (1990) *Justice and the Politics of Difference*, Princeton, NJ: Princeton University Press.

6 Gender and land use in KwaZulu-Natal, South Africa

A qualitative methodological approach

Humayrah Bassa, Urmilla Bob and Suveshnee Munien

Introduction

Land remains a major concern in Africa, particularly in peri-urban and rural areas where land-related inequalities, conflicts and exploitation continue (Rugege *et al.* 2008). In South Africa, unresolved historical inequalities and imbalances linked to colonial and apartheid legacies exacerbate these issues. Furthermore, these processes are steeped in socio-economic and cultural traditions and practices that are highly patriarchal. Persistent discriminatory practices that disadvantage women in terms of access to, as well as control and ownership of, land in South Africa have been generally well-documented in academic literature (Bob 2008; Claassens 2007; Cross and Hornby 2002; Jacobs *et al.* 2011; Meer 1997; Walker 2009). However, most of this research is largely theoretical or policy-orientated and lacks empirically-based research drawn from primary data. One exception is the Jacobs *et al.* (2011) study that provides a quantitative analysis of gender, land and asset rights. This research focuses on a comparison between men and women to highlight the gendered aspects of access to land in peri-urban communities.

The study is based on a qualitative methodological approach to examine land and gender relations in the peri-urban community of Inanda in eThekwini Municipality/Durban, KwaZulu-Natal. The focus on marginalized communities such as Inanda is important since, as highlighted by Payne (2004), access to land and shelter is a precondition that enables the development and provision of other services and livelihood opportunities. Therefore, land is a key resource to secure sustainable livelihoods and support poverty reduction efforts.

Collecting both qualitative and quantitative primary data informs policy debates and program implementation and review. Furthermore, research findings can reveal what appropriate methodological approaches are needed to better understand gender and land relations. This chapter examines land access, use and control among men and women using qualitative data techniques that include mental mapping with Participatory Geographic Information Systems (PGIS), problem-ranking exercises and Venn diagrams, conducted during two focus group discussions (one with men only and one with women only). These methods permit a gender-based comparison of the collected data to analyse land and gender issues.

This chapter is organized into five sections. In the literature review, we summarize key issues relating to gender and land with a focus on the South African context. This is followed by an examination of the methodological approach to research on land, gender and peri-urbanization using qualitative methods. The methodological discussion includes the background that informed the case study and the problems experienced during fieldwork. The next section presents and discusses the research results in relation to land use, rights and institutional dynamics in the study community of Inanda. The conclusion examines how this research, drawn from participatory methodological approaches, contributes to our understanding of gender and land concerns. The results reveal that socio-economic factors such as age, patriarchal practices, marital status and livelihood options influence the nature of gendered relations (which tend to discriminate against women) to land in Inanda. These factors also impact on the distribution and use, as well as the amount and quality, of land available to different households and for communal use. Case studies and qualitative research in particular are shown to be effective methodological approaches to examine gender and land relations.

Gender and land use in South Africa

Studies of gender and land relations in developing contexts, and in South Africa more specifically, highlight several issues that impact unequal access to this resource (Bob 2008; Budlender and Alma 2011; Claassens 2007; Cross and Hornby 2002; Deere and Leon 2003; Rao 2006; Rugege *et al.* 2008; Walker 2009). These issues include historical processes and legacies (specifically colonization and apartheid in South Africa) that affect current patterns of land use and the amount of land availability. In addition, household needs and requirements (including livelihood strategies and activities) are embedded in power relations, especially at community and household levels in traditional authority areas. Patriarchy also emerges as a pervasive issue that impacts the social status of women through male preference in inheritance. Finally, environmental conditions, as well as types of available land, influence land suitability in relation to livelihood activities and the type of ownership or tenure arrangements. These property rights are in turn informed by legislation, policy and cultural practices. As demonstrated by Rugege *et al.* (2008) and Jacobs *et al.* (2011), socio-economic and geographic factors (such as those highlighted here) need to be considered when examining access to, and rights over, land among both men and women in South Africa.

The role of the chieftaincy, or traditional authority, impacts gender and land relations in the South African context. According to Bob (2000) and Rugege *et al.* (2008), the chieftaincy plays a critical role in allocating and managing land in rural and parts of peri-urban communities in KwaZulu-Natal. Therefore, analyzing the relationship between gender and land in areas affected by this form of leadership necessitates an understanding of the chieftaincy. As an institution and stakeholder in land allocation, management and control processes steeped in patriarchal relations, the chieftancy continues to be questioned in these analyses. For example, Budlender and Alma (2011) and Cousins (1996) assert that the control over and allocation of access and use of land have been the primary means by which the chieftaincy has maintained power. Furthermore, "control over land-allocation constitutes the fundamental material basis of the power of the chiefs, and is also the most crucial mechanism for the interplay of corruption and control" (Levin and Mkhabela 1997: 157).

The controversial role of the chieftaincy in relation to land allocation and control in traditional areas remains largely unresolved despite land reform policies that claim to target women and ensure that gender-based discriminatory practices are removed. Budlender *et al.* (2011) in particular make the distinction between rights and customs, which are embedded in cultural systems. Confusion regarding the role of traditional leaders results in uneasiness as well as questions about the effectiveness of the land reform process (Rugege *et al.* 2008). Thus, land reform policy in South Africa is criticized for not setting adequate parameters for a discussion on the impact of customary law and practice on gender equality. Several researchers provide a gendered critique of land reform programs that refer to beneficiaries as "households" because women are often disempowered at the household level (Claassens 2007; Cross and Hornby 2002; Jacobs *et al.* 2011).

The provision of basic needs as well as livelihoods is inextricably linked to access to land. Women's vulnerability to poverty and marginalization are enhanced in contexts where their formal rights to land are limited or when formal rights exist but cannot be exercised due to traditional practices. For example, Rugege *et al.* (2008) illustrate that although they are entitled to land, widows in South Africa are often denied decision-making powers, which remain vested in the hands of male relatives or (in many cases) family members of the deceased husband. These authors indicate that the relationship between gender and property relations should include an examination of both the distribution of property in terms of ownership as well as control over land. Gender equality in legal access to property ownership does not guarantee control over land. This inequality is reinforced by Claassens (2007) and Cross and Hornby (2002) who argue that during apartheid, access and rights to land were largely confined to male heads of households, and where women had access to or rights over land, these were largely mediated through a male relative. To a large extent this pattern continues. According to Hansen *et al.* (2005), marriage and male-to-male inheritance patterns are key components of land tenure that influence how individuals acquire land and resources. Toulmin (2009) also claims that women continue to be denied the right to land tenure, even in areas where they live and raise their families.

In sum, South Africa exhibits many problems pertaining to the equitable distribution of land and gender issues. This study draws from diverse theoretical frameworks and qualitative approaches to analyze the impact of unequal access to, and control over, land rights among women and men in a peri-urban community. The next section briefly examines the importance of qualitative approaches and summarizes the specific techniques used in this study.

Researching land, gender and peri-urbanization using qualitative methodologies

Inanda is a peri-urban community located in eThekwini Municipality, 24 km north of the Central Business District of Durban on the east coast of South Africa (Figure 6.1). eThekwini is the largest municipality in KwaZulu-Natal province and the third largest municipality in South Africa. In 2011, 19.8 percent of South Africans resided in KwaZulu-Natal, which occupies approximately 8 percent of the country's land mass (Statistics South Africa 2012). Furthermore, 86.8 percent of the province's population is of African descent, and 53 percent is female. Inanda is similar to the province of KwaZulu-Natal in that many areas are communal and controlled by traditional authority systems. Additionally, 42.9 percent of households in the province are headed by females (Statistics South Africa 2004).

Large inequities exist within both the built-up sections of Inanda, which experience high levels of development, and the rural areas of this community. In this area, 77 percent of households have a supply of electricity, while the remaining households located in rural areas use fuelwood, paraffin, gas, candles and other sources of energy (Statistics South Africa 2001). Additionally, more males (131,527) than females (91,215) are employed in Inanda, and more women (43,211) are housewives or homemakers compared with males (645) (Statistics South Africa 2001).[1] These gender divisions of labor reflect patriarchal relations within the community. Additionally, 44.6 percent of households have an average annual income of R9,600 (US$1,200) or less (Statistics South Africa 2001). Fewer than half of the households in this community survive on minimal incomes.

The fieldwork in South Africa is grounded in qualitative research that focuses on underlying meanings, experiences and concerns in the social construction of knowledge (von Maltzahn and van der Riet 2006); its focus on the community of Inanda unpacks the tension between qualitative and quantitative research and creates space for the exploration of varied perspectives and voices. As noted by Duraiappah *et al.* (2005), qualitative participatory methodologies reflect decision-makers' efforts to incorporate the perspectives and priorities of local people and their knowledge in policy development and decision-making. This project employs three participatory techniques that are not widely used in research on gender and land issues in South Africa. In particular, mental mapping with PGIS, problem ranking and Venn diagrams are used here as important visualization and diagramming methods. These techniques highlight the significance of livelihoods and the fact that control of and access to land are often gendered and contested.

Discussions with two focus groups (one with fourteen women and one with twelve men) were conducted in this project. The participants for each focus group were purposely chosen to ensure social differentiation in relation to age, position in the community (specifically in relation to community organizational structures), income groups and ownership of land and to highlight the differences between groups in terms of experiences, attitudes and concerns. Tewksbury (2009: 47) asserts that focus groups are regarded as group interviews with "guided conversations in which a researcher (or research team) meets with a collection of similarly

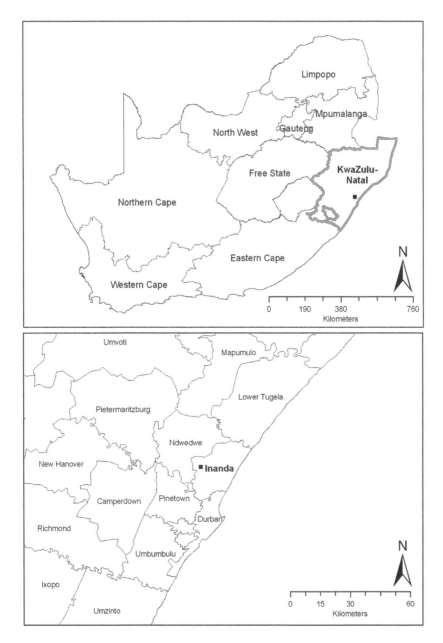

Figure 6.1 Map of South African study area—Inanda, KwaZulu-Natal (2010).

situated persons for purposes of uncovering information about a topic" (2009: 47). Furthermore, de Vos *et al.* (2005) argue that the advantages of focus groups include the ability to observe and to stimulate a large amount of interaction on a topic in a limited period of time by providing rich information and direct evidence regarding similarities and differences in participants' opinions and experiences.

An additional technique in this project includes ranking exercises, or matrices, that determine the order in which objects, concepts and/or resources are deemed to have the greatest importance by participants during a process of consensus-building (von Maltzahn and van der Riet 2006). This technique is useful in identifying issues of concern and prioritizing these issues in order to understand social dynamics and differentiation among women. Pairwise ranking and scoring were used as part of this technique to identify development issues in the community and to establish key barriers to land access and use. Major problems or issues identified by the group were then entered into a matrix, where each problem was weighted against the other before being scored and ranked.

This project also used resource mapping, integrating PGIS, in which people refer to large maps in order to gather information about both social and natural resources of the area. This type of mapping is a graphic participatory technique, which provides physical background as well as community perceptions and socio-economic information about the study site (von Maltzahn and van der Riet 2006). Furthermore, Quan *et al.* (2001) state that participatory mapping with drawn maps can illustrate spatial features and issues of significance that include natural resources, land and/or social resources. The mapping process also creates an opportunity for participants to discuss pertinent issues related to the information being mapped and to share differing opinions and perceptions. PGIS does not attempt to map "factual" information spatially. As von Maltzahn and van der Riet argue, the main contribution of the technique is not in the level of accuracy, but in creating opportunities to examine "what people draw, in what order, in what detail and with what comments" (2006: 123).

Gendered resource mapping permits a comparison of differences in contexts where women's spaces frequently occur between and within lands that are controlled by men (Rochealeau *et al.* 1995). Also, as Kesby (2000) demonstrates, the tactical nature of diagramming permits the contribution of less dominant personalities by allowing participant voices to be heard and to see the results of the research. In addition, Tripathi and Bhattarya (2004) assert that PGIS enables gender empowerment, since ownership of land is a source of social power.

In the mapping exercise, a base sketch map of Inanda, depicting key features in the community such as rivers, main roads, dams, hospitals and schools, was derived from topographic maps and orthophotos of the area. These maps were used to orientate participants and to encourage them to add features, such as key land uses. Additionally, participants were asked to include changes that they would like to see in the community in five years. The maps were sketched on different layers of tracing paper and overlaid onto the base map.

Venn diagrams, another method used in this study, graphically represent the role of institutions and individuals within a community. This visual method permits participants to illustrate social hierarchies and decision-making structures and processes in the community. Venn diagrams include overlapping shapes (circles), which illustrate and summarise relationships, conflicts and issues among different stakeholders using different sizes of circles. The larger the circle, the greater the perceived influence of the institution or organization. In addition, overlapping circles suggest that the institutions or structures are linked in terms of membership and/or decision-making.

Participatory approaches to research such as this are not without their challenges. Transportation barriers were removed by providing a stipend to participants to cover costs and time. Male- and female-only focus groups were conducted to address some of the gender-related barriers identified above. All participants were informed that their comments were confidential and anonymous. Facilitators were conversant in both English and isiZulu, as well as trained in participatory techniques. The reliance on focus groups to generate maps encouraged consensus-building, where the facilitator is crucial to participation and outcomes. Since the group had to agree on the ranking of the problems as well as the jointly formulated Venn diagrams and maps, the building of a consensus proved important to the techniques used in this study.

However, as indicated in the discussions, differences emerged within the groups. Duraiappah *et al.* (2005) note that not everyone within the community is likely to participate due to time constraints, accessibility issues and gender dynamics. For example, some women may be stopped by male relatives or partners from participating. When they are present, men tend to dominate and influence the nature and extent of women's participation. These barriers to participation must be understood and circumvented where possible.

Land use rights and institutional dynamics in Inanda

This research explores important gendered aspects of land use and control in the case study community of Inanda. During the discussions and mapping exercises, participants identified a variety of land uses such as: residential, municipal and government infrastructure and services, as well as commercial, religious, agricultural and natural resource areas. Findings from these participatory methods also differed from the quantitative survey results, where participants identified mainly residential, agricultural and municipal land uses. Similar responses were found for both men and women, although women identified more natural resource and agricultural areas than men. This difference can be attributed to women's overall responsibility for the collection of natural resources such as fuelwood, water and thatch, as well as their involvement in communal gardening. The focus on the importance of natural resources to productive and reproductive strategies in rural areas by Leyk *et al.* (2012) and Shackleton *et al.* (2008) underscore the need to provide a gendered analysis of land use.

This research also demonstrates that women's and men's different uses of land and natural resources have implications for workloads as well as roles and responsibilities at the community and household levels. As one female respondent stated, "fuelwood, medicinal plants and other natural resources have been dwindling in the community. This makes women more vulnerable and they have to work harder to collect or buy these resources which could be found easily in the past." Thus, access to resources impact vulnerable groups such as women in peri-urban communities.

Within the household, the main land uses mentioned by the participants were homes (rented and owned), small garden plots, smallholder commercial agricultural production (mainly sugar cane and chicken production) and small businesses (spaza shops and *shebeens* or local bars). Although similar descriptions of land use within households emerged in the male and the female focus groups, the gendered nature of business activities became evident. Men use land for high-income businesses, such as cattle and goat rearing, renting property, car washes, spray-painting facilities and furniture building, and women usually engage in craft-making, crops and chicken production and preparation of food for sale. According to one female respondent, "when there is money to be made men generally use the land while we (women) do the work." However, a few women run spaza shops and *shebeens* in the community to generate income.

In terms of land control and ownership, both focus groups claimed that land predominantly belonged to the eThekwini Municipality, the tribal chiefs (*amakhosi*), church organizations and individuals who own residential property, which was either purchased or inherited. According to the participants, at the household level, individuals who own land are usually husbands or fathers. Research by Rugege *et al.* (2008) and Jacobs *et al.* (2011) in South Africa as a whole correlates with our findings that males tend to dominate land control and ownership in Inanda.

In addition, the women's focus group identified female-headed households that own residential land as a post-apartheid trend, with widows and female-headed households generally being targeted for housing and land grants by the government (Budlender *et al.* 2011; Jacobs *et al.* 2011). Although some of the policy initiatives have had positive impacts on women in Inanda, several female participants pointed out that only a few women benefitted and many in similar predicaments were on waiting lists to take advantage of these programs. One respondent stated:

> I put in an application for a house more than three years ago. I try almost every month to find out what is happening but there has been no response. There are many women like me who were encouraged to apply for houses but we are still waiting.

Thus, it is still common in the community for sons to inherit land even when women have been granted land rights.

Patterns of land ownership generally revert to males despite legal changes and land reform that targets women. As Budlender and Alma argue, "Merely passing legislation is of little effort without the necessary resources for implementation, without informing and educating all relevant actors on the provisions of the legislation, without monitoring the reforms, and without effective sanctions on failure to implement" (2011: xii). Younger women expect that their land needs will be met when they marry, a sentiment echoed in the male focus group. However, a significant proportion of households in this area are headed by women, and many women do not marry. These inequities were shared by one respondent in the female focus group who stated, "I already have two children who I am taking care of myself. I work and I take care of my parents yet the land belongs to my father and it will not eventually belong to me." Thus, cultural and patriarchal practices and attitudes persist despite considerable changes in local and national policies as well as women's increased role as primary breadwinners in households. These results reinforce other studies that indicate that land reform and legal changes are insufficient to address gender discriminatory practices (Bob 2008; Claassens 2007; Jacobs *et al.* 2011).

Problems faced by the female and male participants of this study, with regard to land, are depicted in the pairwise ranking matrices in Table 6.1. One (1) is ranked as the most important problem and ten (10) is deemed to be the least important among those identified during the focus group discussions. Results from the ranking exercises indicate that both women and men identify and rank problems differently. For example, women perceive the top four land-related problems to be the high price of land, lack of available land to purchase, difficulty accessing water and land theft and robbery (Table 6.1). In contrast, the top three problems identified by men are lack of facilities and infrastructure, land being used for housing and insufficient land or unavailable land. Theft is a problem identified by women but not by men, indicating that women are more vulnerable to crime in communities. Also, women claim that people do not pay them rent, while men did

Table 6.1 Ranking exercise on land issues in Inanda

	Women	Men
Conflict occurs over the land	6	10
Crime is a problem due to theft and robberies	3	—
Decreasing land size and availability	—	4
Difficult to access land	—	4
Difficult to access sufficient water	3	—
Difficult to maintain the land	8	—
Difficulty in making paying rates	8	7
Insufficient land/land is not available	—	3
Lack of facilities and infrastructure	—	1
Land inheritance/who should inherit land	—	9
Land is not available to buy	2	—
Most land is used for housing	—	2
People don't pay women rent	5	—
Poor quality of soil	10	6
The price of land is too high	1	—
Tools and equipment are not available	7	—
Water table is too high	—	7

Source: Bassa, Bob and Munien, authors' fieldwork 2008–9.

not experience this problem. This is a form of theft and indicates that as female landowners, they are not respected.

The cost of land and agricultural inputs appears to make it difficult for women to make productive use of land if they have access to it. For men, the main problems are the amount of land available (and not necessarily the ability to access the land) and the provision of services and infrastructure. In terms of land availability, Woodhouse (2003) asserts that land resources in rural and peri-urban areas are unable to keep up with growing pressures due to high population densities, which are linked to their status as former homelands, and population growth. Consequently, large numbers of people share resources from a decreasing land base.

In the ranking exercise, men identified land inheritance decisions as a problem—primarily because male sons tend to migrate out of Inanda to find work. However, the respondents did not account for females or daughters in these land inheritance situations. The fact that both men and women identified conflicts over land supports other studies that identify tension over land and related resources as key problems in South Africa (Bob 2010; Rugege *et al.* 2008). In this study, conflicts among family members in particular were raised by women. One such conflict occurred in situations where women owned land but were not allowed to make decisions about how the land should be used. As one respondent stated, "My friend owns the land she lives on but her husband's family keeps interfering in household matters. She lives with her two daughters and they constantly fight with relatives to leave them alone." Finally, while quality issues pertaining to land were raised by both men and women, and the high water table issue was raised by men, these were not ranked as serious problems. It does indicate, however, that the quality of soil in Inanda may be limiting productive livelihood options, contributing to the poverty experienced by many households.

Maps obtained from PGIS exercises depicting land ownership among the different groups reveal important gender differences in perceptions of land and its uses. As illustrated in Figure 6.2, women in Inanda are aware of community gardens, shopping centers (such as Bridge City and IDUBE Village) religious sites, taxi ranks, schools, clinics and police stations. They also identify land owned by the chief and categorize housing settlements into low-cost, formal and informal housing settlements. This PGIS map suggests that since they are aware of community gardens within Inanda, female participants perceive them to be important. The significance of these community gardens is further noted by the desire among these women to have more of them in the future. This awareness is partly due to the presence of two women who were involved with community gardens. As one of these women stated, "Community gardens are an important source of income for many women. While we started out producing enough crops just for our families, we now grow more than we need and can sell vegetables to earn money." The map also suggests that in the future, women wish to have more community facilities such as clinics, a swimming pool and old age homes. Moreover, the women also indicated that they would like to have more municipal offices in Inanda for

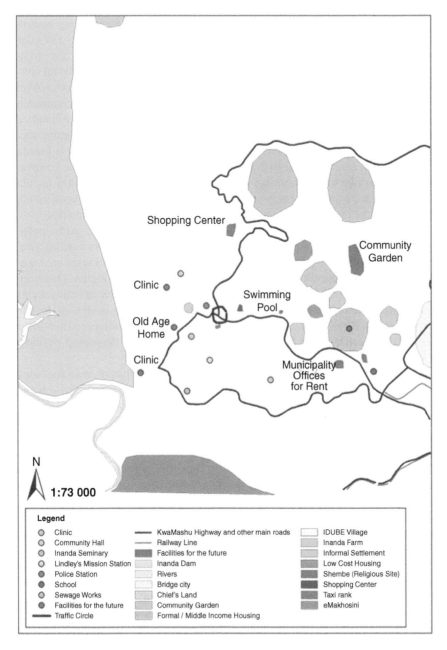

Figure 6.2 Women's perceptions regarding current and future land use in Inanda. (Only the base features are to scale; fieldwork 2008/2009.)

small business and other non-agricultural activities. Thus, there is a strong desire to engage in income-generating activities among women in Inanda.

In contrast, men in Inanda identified soccer fields, clinics, shopping centers, cemeteries/burial sites, churches, schools, community gardens, forests and quarries as well as IDUBE and eBohleni Villages (Figure 6.3). They were also aware of townships and informal settlements, and although they indicated several community gardens in Inanda, the location of these gardens is incorrect. This mistaken location could be due to the fact that none of the men are involved in community garden activities while, as indicated earlier, two of the women are active members of these community gardens. The PGIS map also shows that in the future, men would like more hospitals, sports fields, libraries and municipal offices in Inanda. These aspirations are similar to those expressed by female participants; however, males would like to see more key business nodal developments within Inanda. Overall, men are more business-orientated than women, who only identified one location for the municipality to provide offices for small business ventures.

Figure 6.4 is a map of actual land use in Inanda compared to participants' perceptions of how land is used. Whilst the PGIS mapping exercises were informative, there are limits to participatory mapping. For example, PGIS maps are based on perception and are not necessarily accurate descriptions of a region. Therefore, regions that appeared to be "empty" in the perception maps are, in reality, being used. Furthermore, the actual scarcity of land in Inanda is not evident in the perception maps. Finally, since perception maps are neither geo-referenced nor drawn to scale, the activities and features are not always located in the correct positions (Tripathi and Bhattarya 2004).

The main features on the maps are participants' perceptions of main land uses, specifically aspects such as community gardens, forests, plantations and schools. The results from this map, and comparisons with previous figures, reveal that the focus groups were unable to identify many of the land uses that actually exist in the community, especially in terms of specific location and/or extent. Furthermore, there was a tendency to exaggerate the spatial extent of land uses that they are familiar with (i.e. the community gardens). Therefore, it is important to note that while mental mapping exercises are useful, complementing the results with geo-referenced data can provide more detailed information of land uses. Moreover, the contradictions reveal how people perceive resources and land uses and particularly how knowledge is biased towards land uses with which respondents are familiar. In contrast, results show that geo-referenced data are often at a scale that omits land features that are visible at a finer resolution and may impact on livelihoods and people's spatial experiences. For example, results show that forests identified by respondents were categorized as rural and agricultural land in formal maps.

The Venn diagrams in Figures 6.5 and 6.6 illustrate the gender ratio of members in institutions and organizations and their relative position in the community. Venn diagrams are visual methods used to represent the roles of institutions within a community. The larger the circles, the greater the importance of these institutions as perceived by the participants. The overlapping circles indicate that members belong to multiple institutions. The higher position on the chart correlates to

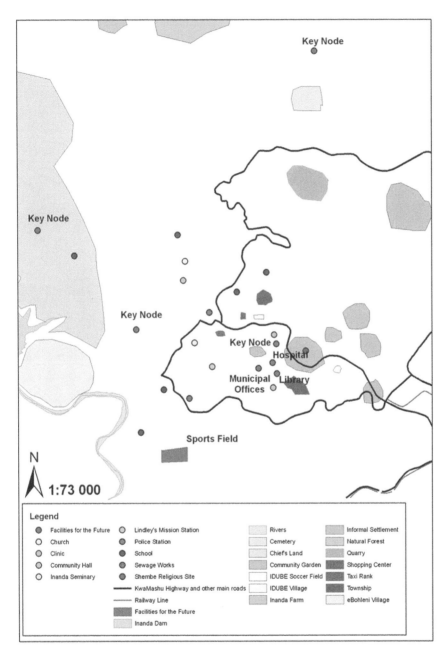

Figure 6.3 Men's perceptions regarding current and future land use in Inanda. (Only the base features are to scale; fieldwork 2008/2009.)

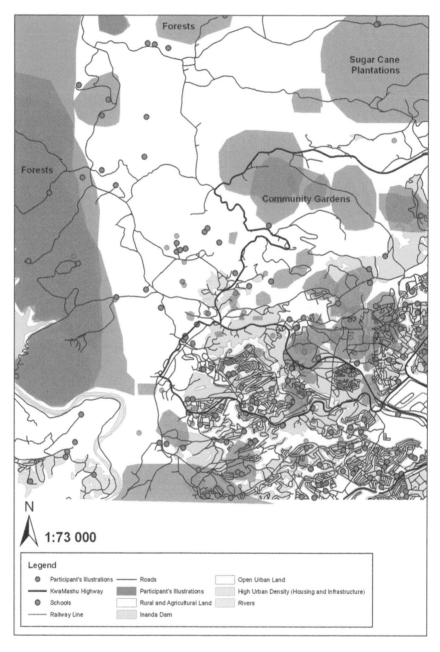

Forests

Sugar Cane
Plantations

Forests

Community Gardens

N

1:73 000

Legend

⊙ Participant's Illustrations	—— Roads	☐ Open Urban Land	
—— KwaMashu Highway	▓ Participant's Illustrations	▓ High Urban Density (Housing and Infrastructure)	
⊙ Schools	☐ Rural and Agricultural Land	▓ Rivers	
—— Railway Line	▓ Inanda Dam		

Figure 6.4 Actual land use compared to participant perceptions of land use in Inanda.
(Fieldwork, 2008/2009.)

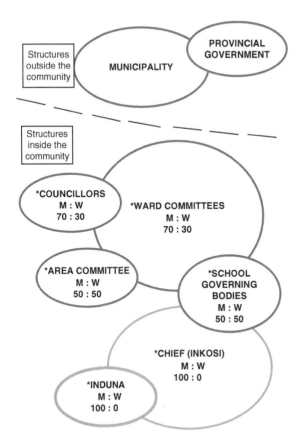

Figure 6.5 Women's group Venn diagram. (Ratios are in percentages; M = men and W = women; fieldwork 2008/2009.)

the perception that the institution is more powerful. In this study, men represent the majority of members and hold most of the leadership positions within these organizations. With the exception of school governing bodies, which are comprised of mainly women, most institutions are entirely dominated by men, with a few equally represented by men and women. According to Rugege *et al.* (2008), women still believe patriarchy persists within traditional leadership structures and the chief and local *Induna* continue to be male-dominated. Male dominance in traditional structures is also evident in the male Venn diagram (Figure 6.6). Furthermore, women tend to have greater confidence in governmental organizations, such as the municipality and ward committees, which are democratically elected. These circles are larger than other organizations on the diagram. In Inanda, traditional structures are almost exclusively the domain of males.

Thus, Venn diagrams of community structures and institutions are useful in understanding power dynamics within these communities. The visual depictions

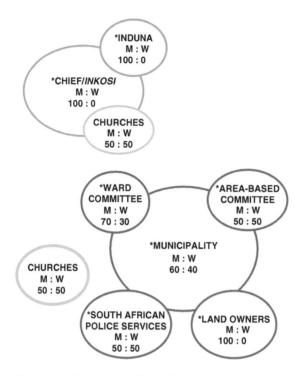

Figure 6.6 Men's group Venn diagram. (Ratios are in percentages; M = men and W = women; fieldwork 2008/2009.)

are also useful to participants who, in both focus groups, wanted to keep the charts they generated so that they could "show" other community members and leaders the landscape and different structures in the community as they see them. Thus, the use of qualitative research approaches (such as those used in this project) move away from extractive data collection to more innovative, interactive engagement with participants.

Conclusion

This research identifies several gender-related issues that address access to, and control of, land in a peri-urban community of South Africa. Socio-economic factors such as age, marital status and livelihood options influence the gendered distribution and use of land in this community. In the Inanda case study, key aspects of gender and land relations include cultural practices embedded in patriarchal institutions, the amount and quality of land available in communities, available assets to increase productive use of land and policy and program changes. This study outlines discriminatory practices against women, such as limited access to land and decisions being made by male relatives. Consequently, many women are denied access to and control over land and related resources. For example, in cases where women have ownership of land, preference for male inheritance results in land reverting to male hands.

This research also highlights the importance of using qualitative approaches to examine gender and land relations in peri-urban communities and particularly underlines how access to land in post-apartheid South Africa is gendered at both the community and household levels. Case studies and qualitative methods are effective in ascertaining how gender plays a role in relation to land access, use and control. The research demonstrates that diagrams and ranking exercises are innovative ways of unpacking which issues are important to men and which are important to women. The gendered nature of land use is evident in several of the participatory techniques, particularly the Venn diagrams, which show how decision-making structures in Inanda are mainly male-dominated. Additionally, the ranking and mapping exercises indicate that men and women perceive and experience land within Inanda differently, however, common concerns and problems are also discernible.

Discussions linked to the exercises also allow one to develop future research projects that build on gender and land issues in the context of South Africa. For example, differences and commonalities emerge that can inform policy developments and efforts to address gender inequalities in relation to land access and control. The primary land uses for both men and women are residential and agricultural production. Additionally, both men and women would like more development of infrastructure in the area. Land-related conflicts are also raised as problems by both men and women; however, there are clear differences in relation to problems experienced by men and women. On one hand, women generally identify concerns that relate to their ability to acquire and purchase land and exercise rights over their own land. Men, on the other hand, raise concerns about accessing land, the amount and type of land available and the lack of facilities and infrastructure. Thus, qualitative research techniques are effective in understanding gender and land issues in peri-urban communities in South Africa and other developing contexts as well.

Note

1 At the time of publication, the 2011 South African Census aggregated data has only been released at the national level and not at the provincial level. The latest data at the provincial level are the 2001 statistics.

References

Bob, U. (2000) "Gender, the role of the chieftaincy and rural women's access to land under the land reform program in KwaZulu-Natal," *Alternation* 7(2): 48–66.

Bob, U. (2008) "Rural women's relation to land resources in KwaZulu-Natal: issues of access and control," *Alternation*, 15(1): 110–34.

Bob, U. (2010) "Land-related conflicts in sub-Saharan Africa," *African Journal on Conflict Resolution* 10(2): 49–64.

Budlender, D. and Alma, E. (2011) *In Focus: Women and Land-Securing Rights for Better Lives*, Ottawa: International Development Research Center.

Budlender, D., Mgweba, S., Motsepe, K. and Williams, L. (2011) *Women, Land and Customary Law*, Johannesburg: Community Agency for Social Enquiry.

Buregeya, A., Garling, M., Craig, J. and Harrel-Bond, B. (2001) *Women's Land and Property Rights in Situations of Conflict and Reconstruction*, United Nations Development Fund for Women.

Claassens, A. (2007) "Women and land," Draft document for discussion at the PFOTA Workshop.

Cousins, B. (1996) "Livestock production and common property struggles in South Africa's agrarian reform," *The Journal of Peasant Studies*, Special Issue on the agrarian question in South Africa 23(2 and 3): 166–208.

Cross, C. and Hornby, D. (2002) "Opportunities and obstacles to women's land access in South Africa," a research report for the Promoting Women's Access to Land Program, University of the Witwatersrand.

Deere, C. D. and Leon, M. (2003) "The gender asset gap: land in Latin America," *World Development* 31(6): 925–47.

de Vos, A. S., Strydom, H., Fouche, C. B. and Delport, C. S. L. (2005) *Research at Grassroots for the Social Sciences and Human Service Provisions*, 3rd edn, Pretoria: Van Schaik Publishers.

Duraiappah, A. K., Roddy, P. and Parry, J. (2005) *Have Participatory Approaches Increased Capabilities?* Winnipeg: International Institute for Sustainable Development.

Hansen, J. D., Luckert, M. K., Minae, S. and Place, F. (2005) "Tree planting under customary tenure systems in Malawi: impacts of marriage and inheritance patterns," *Agricultural Systems* 84: 99–118.

Jacobs, K., Namy, S., Kes, A., Bob, U. and Moodley, V. (2011) *Gender Differences in Asset Rights in KwaZulu-Natal, South Africa*, Washington, DC: International Center for Research on Women.

Kesby, M. (2000) "Participatory diagramming: deploying qualitative methods through an action research epistemology'," *Area* 32 (4): 423–35.

Levin, R. and Mkhabela, S. (1997) "The chieftaincy, land allocation, and democracy," in R. Levin and D. Weiner (eds) *No More Tears: Struggles for Land in Mpumalanga, South Africa*, Trenton, NJ: Africa World Press, pp. 153–73.

Leyk, S., Maclaurin, G. J., Hunter, L. M., Nawrotzki, R., Twine, M., Collinson, M. and Erasmus, B. (2012) "Spatially and temporally varying associations between temporary outmigration and natural resource availability in resource-dependent rural communities in South Africa: a modeling framework," *Applied Geography* 34: 559–68.

Meer, S. (ed.) (1997) *Women, Land and Authority: Perspectives from South Africa*, Braamfontein: National Land Committee.

Payne, G. (2004) "Land tenure and property rights: an introduction," *Habitat International*, 28: 167–79.

Quan, J., Oudwater, N., Pender, J. and Martin, A. (2001) *GIS and Participatory Approaches in Natural Resources Research: Socio-economic Methodologies for Natural Resources Research – Best Practice Guidelines*, Chatham, UK: Natural Resources Institute.

Rao, N. (2006) "Land rights, gender equality and household food security: exploring the conceptual links in the case of India," *Food Policy* 31: 180–93.

Rochealeau, D., Thomas-Slayter, B. and Edmunds, B. (1995) "Gendered resource mapping: focusing on women's spaces in the landscape," *Cultural Survival Quarterly* Winter: 62–68.

Rugege, D., Bob, U., Moodley, V., Mtshali, S., Mtunga, O. and Mthembu, A. (2008) "A baseline survey on the Communal Land Rights Act in KwaZulu-Natal report," Communal Land Rights Act Baseline Survey, Department of Land Affairs.

Shackleton, S., Campbell, B., Lotz-Sisitka, H. and Shackelton, C. (2008) "Links between local trade in natural products, livelihoods and poverty alleviation in a semi-arid region of South Africa," *World Development* 36(3): 505–26.

Statistics South Africa (2001) *Census in Brief*, Pretoria: Statistics South Africa.

Statistics South Africa (2004) *Provincial Profile 2004: KwaZulu-Natal*, Pretoria: Statistics South Africa.

Statistics South Africa (2012) *Census 2011: Statistical Release*, Pretoria: Statistics South Africa.

Tewksbury, R. (2009) "Qualitative and quantitative methods: understanding why qualitative methods are superior for criminology and criminal justice," *Journal of Theoretical and Philosophical Criminology* 1(1): 38–58.

Toulmin, C. (2009) "Securing land and property rights in sub-Saharan Africa: the role of local institutions," *Land Use Policy* 26(1): 10–19.

Tripathi, N. and Bhattarya, S. (2004) "Integrating indigenous knowledge and GIS for participatory natural resource management: state of the practice," *The Electonic Journal of Information Systems in Developing Countries* 17(3): 1–13.

von Maltzahn, R. and van der Riet, M. (2006) "A critical reflection on participatory methods as an alternative mode of enquiry," *New Voices in Psychology* 2(1): 108–28.

Walker, C. (2009) "Elusive equality: women, property rights and land reform in South Africa," *South African Journal on Human Rights* 25: 467–90.

Woodhouse, P. (2003) "African enclosures: a default mode of development," *World Development* 31(10): 1705–20.

7 Participatory mapping of women's daily lives

Perspectives from rural Uganda

Deborah Naybor and Ram Alagan

Introduction

During the past few decades, studies have highlighted gender equality, representation and identity as important themes for the empowerment of women (Barbaraj
2004; Ellis *et al.* 2006; Momsen 2006). Information technology has the potential
to engage women in development, which in turn improves donors' and policymakers' understanding of gender and increases access to local knowledge for
better decision-making on all levels. In order to become fully invested in the
development process, women need access to information technology as well as
inclusion in the processes of planning, obtaining, processing and interpreting data
that affects their lives. The International Center for Research on Women (ICRW)
noted that "women are often seen only as 'users' or 'receivers' of technology, not
as innovators" (Gill *et al.* 2010: 7). The ICRW contends that in order to enable
them to use technology for economic advancement, women need to be involved
in the design, deployment of and access to technology. This chapter examines the
use of participatory research methods that seek to explain rural women's time use
in Uganda and examine the impact of their spatial and temporal patterns on all
levels of human development.

Developing a comprehensive understanding of women's daily lives in remote
rural villages is difficult given the multiple constraints on the research process. For
instance, the presence of the researchers influences normal time use and disrupts
daily routines. Thus, in order to understand gendered time use and movement, the
influence of outsiders should be reduced or eliminated, while methods of mapping mobility must be improved. In addition, the self-reporting of time use is
often incomplete, especially among cultures who have little use for chronological
measurement and among illiterate populations (Antonopoulos and Hirway 2010;
Abdourahman 2010). Therefore, accumulating detailed knowledge of marginalized women within normalized productive and reproductive labor requires the
development of new methodologies regarding remote data collection methods that
depict temporal and spatial relationships.

This study uses Participatory Geographic Information Science (PGIS)[1] and,
specifically, Global Positioning System (GPS) data trackers to monitor temporal and spatial locations in the daily mobility patterns of twenty-seven women in

rural Uganda. The information gathered from this fieldwork provides insight into constraints and challenges faced by the group as they work to provide for their families. Although outwardly the group members appear to experience the same levels of poverty, this study shows that there are subtle differences in the development levels and time use of individuals that are not apparent without participatory input and the use of geospatial technology. This study finds that important differences in the patterns of movement are often based on the economic conditions of participants. Traveling to market for those who are economically stable may be to purchase food, whereas for others, going to the market may mean begging for enough food to provide one meal for the day. The analysis illustrates the strengths, complexities and challenges of participatory geospatial research, while exploring the relevance of locational data and qualitative information on women farmers in rural Uganda. The project also demonstrates the potential use of these spatial technologies and participatory mapping techniques for evaluating gender dimensions of the everyday life of rural women.

Walker and Shalini (2009) note that gender aspects of spatial exclusion are a vital component of social marginalization and vulnerability. Use of contemporary mapping and analysis software highlights the geospatial and temporal entrapments that are associated with poor women's access to information and full participation within their society. The quantitative data provided through this study relies upon the full cooperation and participation of the women. This chapter also underlines some of the difficulties in implementing Geospatial Spatial Information Technology (GSIT), but opens the way for further investigation into what constitutes the best practices for avoiding gender-biased, top-down decision-making in the fields of development studies and feminist geography.

Gender, information technology and GIS

In order to understand the influence of space and time on the capability of women in rural Uganda, this work draws from studies of spatial entrapment (Hanson and Pratt 1991), time poverty (Blackden and Wodon 2006; Lawson 2008; Robles 2010) and the gendered use of time (Spring 2000; Robertson 1984; Mook 1976; Boserup 1970). Some feminist scholars maintain that development discourse is largely based on the assumption that women's lives have not changed significantly during the past several decades (Kabeer 2005). However, in recent years, development practice has integrated information technology (IT) and gender norms to a greater extent. Properly employed technology assists in improving planning and the strategic deployment of development programs like those that identify areas of need through community participation in GIS mapping.

The use of GIS in the study of gender issues has become an innovative research model among scholars (Gilbert *et al.* 2008; Bosak and Schroeder 2005; Walker and Shalini 2009; Kwan 1999). This chapter explores gendered perceptions within GIS research and addresses the current limitations of such theories and applications. Kwan's seminal work on GIS as a method in feminist geographic research, demonstrates that "when individual level data are available, GIS methods can be attentive to the diversity and differences among individuals" (2002: 651). GIS uses mapping and spatial analysis for in-depth comprehension of the barriers to daily decision-making and mobility. Geospatial data allow for the inclusion of spatial and temporal components in formulating development programs. Incorporating this information enables the design of programs that increase efficiency and incorporate home-based sustainability for rural women in Africa. Kyomuhendo and McIntosh (2006) observe that if geographic mobility includes women, increased economic participation and growth follows.

Early GIS programs were only capable of processing quantitative digital information; however, multimedia GIS[2] now handles a much wider array of representational possibilities than the mere processing of data. This technology provides the tools to better represent data to donors, agencies and participants in a way that is understandable and meaningful without creating top-down, exclusionary data. Alagan (2007) argues that critical GIS debates have shifted this technology toward theoretical arguments in GIS applications, especially through PGIS's emphasis on community involvement. By merging both expert and community knowledge, PGIS lends itself to community empowerment and collaborative decision-making. One of the goals of PGIS is to involve wider participation in development projects and to empower the public in decision-making processes through inclusion and the sharing of tacit knowledge rather than exclusion through technology.

GIS is gaining prominence in the study of poverty issues in sub-Saharan Africa and allows for the geospatial analysis of challenges that impact the poor. For example, GIS provided information for participatory planning of informal settlement (slum) upgrades in South Africa (Abbott 2003). A telling study of poverty by Noor *et al.* (2008) used remotely sensed night-time light levels as a proxy for poverty. Through the comparison of poverty indices and mean night-time brightness, correlation between poverty levels and light creates an inexpensive alternative for identifying areas of asset-based poverty. Yamano and Kijima (2010) use GIS to compare socio-economic data that allow them to analyze road distance and condition from home to market in rural Uganda. They found that perishable and high-income crops may be at risk of damage on long, rough roads, thereby limiting market opportunities for farmers in remote areas. This use of geographical data to analyze the relationships between place and poverty is critical in targeting and customizing aid and development funds. GIS also has the potential to evaluate gender issues through the use of geospatial technology in order to untangle the complexities of barriers and opportunities that exist for women and influence their ability to fully participate in development.

GIS is intrinsically spatial and has great potential to understand the spatial mobility of women and their day-to-day work. Kwan (1999; 2002) has used time geography in GIS to trace and visualize women's life paths in space–time and the impact of space–time rigidity on their daily activities. This research links women's qualitative data (e.g. narrative, stories and interviews) and multimedia visual representations (e.g. photos, graphs, charts, video and voices) within GIS in order to understand their daily space–time paths and to create a way to tell their stories.

Changes in GIS technology and methods move toward an implicitly integrated approach that illustrates the importance of women's engagement in development issues, especially in poverty reduction measures. Although geospatial research has historically enlightened only those with access to technology, it is critical to share this knowledge with poor communities and local leaders. When tracking the movement of rural women using GPS, researchers must consider the extent to which geospatial technology can provide these women with knowledge about their own role in economic development.

Time use and gender

Recent attempts to define poverty and development use multiple indicators of wealth and wellbeing, which have been incorporated into quantifying the human condition. Roy (2010) claims that belief in the abilities of the poor, through programs like microcredit, has provided a more constructive view of poverty, especially in relation to gender. She also notes that the financial success of such loan programs may be deceptive. Development is subject to cultural and social dimensions beyond the simple measurements of goods and services produced and also "attends to the social consequences of production" (Peet and Hartwick 2009: 2). More than just a means to increased consumption, development also includes those hard to quantify intangibles such as quality of life, security, knowledge and control over one's own destiny.

Development scholars note an increased emphasis on non-income measures of poverty and time use data, such as studies of household production, in understanding poverty (Lawson 2008). Time is especially important in the research of poor women, who may have little control or ownership of assets and for whom time is a critical resource. One method of contrasting the availability and use of productive time to non-productive leisure or rest time is the measurement of time poverty. Bardasi and Wodon define time poverty as:

> [W]orking long hours without choice because an individual's household is poor or would be at risk of falling into poverty if the individual reduced her working hours below a certain time-poverty line. Time poverty is thus understood as the lack of enough time for rest and leisure after accounting for the time that has to be spent working, whether in the labor market, doing domestic work, or performing other activities such as fetching water and wood.
> (Bardasi and Wodon 2010: 45)

According to Blackden and Wodon (2006), patterns of time use are not equitably distributed among men and women. Poor women generally do not have sufficient time in the day to earn money through production of goods and services, given their responsibility for reproductive work such as care giving and unpaid household work. Additionally, time spent on chores and childcare relates to spatial mobility and access. For example, Barwell (1996) reports that gathering firewood and other fuels for cooking is a feminized task in the Mbale District in

Eastern Uganda. He estimates that more than 900 hours a year could be saved in transport time if water were within a six-minute walk and firewood within a thirty-minute walk. The use of geospatial tools in time use study provides a visual and analytical tool for understanding how time commitments and spatial connections are intrinsic to poverty.

Time constraints are critical to the success of development policy and program design. Ali's (2009) research found that distance and travel time to work greatly impacts potential earning power, making tools such as GIS appropriate technologies for development research. Incorporating poor women within development programs requires analysis of their location within both space and time, since a woman's ability to earn income and provide for her family is linked to spatial mobility. However, the ability of women to alter their mobility and access information and markets is often constrained by cultural restrictions. Kwan (1999) argues that conventional accessibility measures suffer from an inherent gender bias and instead recommends a space–time feasibility study to eliminate this bias. Time geography thus provides the spatial analysis and modeling for further study of the interaction between space, time, gender and poverty.

Rural women in Uganda's Rakai district

While many observational, ethnographic and time use studies of African women have been conducted (Kyomuhendo and McIntosh 2006; Memis and Antonopoulos 2010; Asiimwe 2010; Serra 2009), the temporal-spatial patterns of poor rural women remain largely unknown in development studies. In order to develop a deeper understanding of the mobility and time use of women in rural Africa, a study of time use and mobility using PGIS, GPS and ethnographic methods was conducted in the village of Nakagongo in the Rakai District of southeastern Uganda (Figure 7.1). In this study, women farmers struggle with insufficient time in many ways. Much of their day is spent gathering wood for cooking, preparing meals and securing food for their families. They are often responsible for subsistence farming for their family and selling their excess crops in the local market. In addition, their waking hours are spent on reproductive, non-income-producing labor, such as caring for the sick, the elderly and small children.

In addition to the capital city, Kampala, Uganda is divided into 111 districts, which are further divided into counties, sub-counties and town councils (Republic of Uganda, Ministry of Local Government); parishes, formed by the predominate Catholic Church, are also a commonly used designation for local governance. The Rakai District, created in 1980 in the western part of the central region of Uganda, has an estimated population of 466,300 and is comprised of three counties, 18 sub-counties and 103 parishes. In total, 96 percent of the district lives in rural settlements, with a population density ranging from 60 to 141 people per square kilometer (Rakai District 2010). The parish of Bethlehem, located within this district, has a population of approximately 3,000 adults, 5,000 youth and 10,000 children. An informal report from a local village leader stated that there are approximately 200 shops within the parish, including the informal market stands

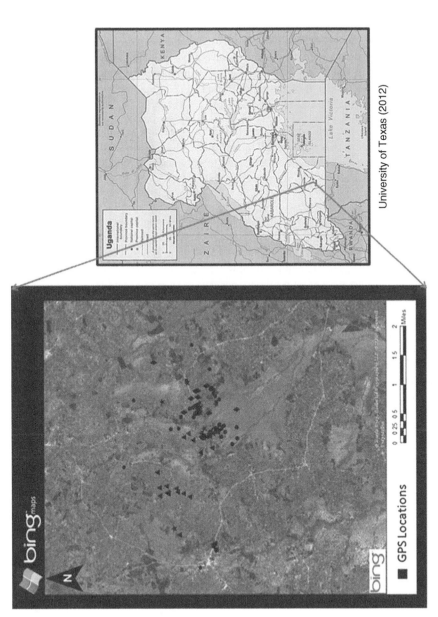

University of Texas (2012)

Figure 7.1 Case study area showing GPS data points. (Bing satellite photos with GIS data; fieldwork by Deborah Naybor, 2012.)

along dirt roads, and about ten formal trading centers (interview Fred Swerwangu, January 7, 2012).

The study area is within Bethlehem parish in the village of Nakagongo, an area made up of a dozen or so small villages, which surround various market centers. The township is estimated to have an adult population of approximately 300, including approximately 200 adult women, aged 18 to 80.

The participatory group for this study consisted of twenty-seven women, aged twenty-three to seventy-three, who work together on small-scale development projects. All of the women have been widowed and about 30 percent have remarried. They live in simple mud or brick homes, without running water or electricity. A thirty-nine-year-old participant reported that:

> My house is too old and it [can] fall at any time and whenever the rainfall [comes], rain water become full in the house. And it is too small for my family and the land we are having is too small to grow enough food and it is infertile which do not allow our crops to grow well.

Mwaka (1993) found that rural women in Eastern Uganda spent approximately 15.5 hours per day on household work and crop production. She notes that women in Uganda are responsible for 80 percent of all food production, and this work is mostly done with inefficient hand tools. The women in this study self-reported that they work an average of 6.6 hours per day in crop production and 5.7 hours per day on reproductive labor such as cooking and caring for children.

Most of the group members have had some schooling, although the average level of education is only four years of primary school. Two-thirds of the women report that they are able to read and write, which is comparable to the 73 percent national literacy rate for females in Uganda (UNICEF 2012). All the women participate in subsistence agriculture, but only 44 percent claim that they have sufficient yields to sell excess crops and nearly 80 percent earn income from sources other than selling food crops. Many of these women raise animals for sale, and several make and sell home-brewed local beer or make crafts, such as baskets or jewelry. One of the most significant indicators of their individual poverty is poor health and nutrition. Two-thirds of the women reported that they are unable to work five days or more a month due to illness, and 40 percent of the group visits the local clinic once a week a more. They report poor diets with little meat, chicken or fish, and the available food is of poor quality and inadequate quantity. In answer to questions about particular challenges, most women reported that poor health is the biggest problem they face. As one of the women noted:

> I am an AIDS victim. My life all the time seems malnourished, no good hope for the future. It seems hard for me to buy a good meal to adjust the diet which extends my life. All the time I lack funds to facilitate my life status. My health status is bending to poor.

This research found that the women who participated in this study dedicate an average of 39.5 hours per week to income-producing work and spend 53.3 hours per week on subsistence and care-giving work. Raising livestock or growing food for market sale increases their level of human development; however, limited non-agricultural work opportunities have no significant impact. On average, they travel six miles per day, mostly on foot, and 40 percent of their travel is within 0.15 miles of their home, indicating spatial fixity. Women with a higher level of education exhibit lower levels of fixity and higher levels of development, which indicates increased opportunities.

Participatory geospatial research: Mapping women's lives

Studying time use among rural African women using geospatial technology requires logistical and technological problem solving. Preparations for this project included group selection, identifying GPS units that would receive satellite signals in the region and resolving short battery life issues. This research involved the identification of hundreds of data points and thousands of paths that provided time use and mobility data for each of the participants. The use of personal interviews and ethnographic observations enabled the identification of time use categories at various locations. In addition, semi-structured questionnaires and interviews provided additional quantitative and qualitative data for analysis and allowed for the creation of a modified human development index for the comparison of factors such as education, health and living standards.

Hand-drawn participatory mapping was completed by the women prior to mapping with geospatial technology (Figure 7.2). Through this exercise, they completed drawings of their homes and important locations throughout the community such as the local well, churches and shops. One participant noted that she had never held a pen in her hand and others worried about drawing the map correctly. The women were concerned about whether they were "doing it right" and often turned to the men (who acted as translators) for help. However, as they gained confidence in their own abilities and were assured that there was no wrong way to draw the map, they added more details like the location of the market or clinic.

Gathering information on women's patterns of movement and time use was undertaken during various periods in 2012–13. Data collection for the study used LandAirSea GPS trackers carried by the participants as they moved around homes or traveled to markets or other areas. A key advantage of GPS tracking is the ability to create active roles for women participants and to obtain data without interference by researchers. The women stated that inclusion in the study provided a chance to "show outsiders how hard we work" and demonstrate their active roles in income-development projects.

Explanations and demonstrations about how the tracking devices worked were critical to the group's sense of inclusion in the research. At each step of the process, a group meeting was held with the women where maps of patterns of movement and the findings from the interviews were shared without identifying

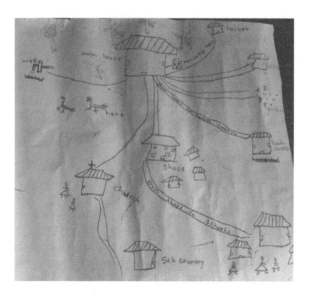

Figure 7.2 Participatory map showing spatial relationship concepts. (Photo by Deborah Naybor, 2012.)

individual participants. The women took great care to answer interview questions and provide information through tours of their homes. Questionnaires were used to gather data about housing, family, work, transportation, health and education. Twenty categories were numerically scored in order to create a modified human development index (MHDI), which measured each individual's relative poverty (Table 7.1). Broad categories of investigation were: living standards, family interaction, health and nutrition, livelihoods, transportation and education. Various components in each category of this index were analyzed for statistically significant relationships between time use and patterns of movement. By re-examining these questions on an annual basis, the measure of development changes can be monitored at a future date.

Gathering tracking data for this study proved more difficult than anticipated, as the women were sometimes too sick to move from their homes or did not carry the trackers. In addition, the battery life of the devices was only a few days, and spare batteries had to be charged using solar chargers due to the absence of electricity in the village. Research assistants were often unable to fully charge the batteries within the short time frame because of other obligations or rainy conditions. In order to protect the privacy of participants, women were given the option of leaving the tracking devices at home on occasions when they did not wish their locations to be known, such as visits to a HIV/AIDS clinic. The Western notion of privacy differs from privacy in this rural village, where large families live in undersized buildings without separate rooms. As one Ugandan woman explained, "what you call privacy, we call loneliness. Eating a meal alone is to go hungry." Although

Table 7.1 Summary of modified Human Development Index variables (as a percentage of participants)

Variable	%
Living standards	
Own their home	63.0
Home has metal roof	66.7
Home has concrete floor	11.1
Home has electricity	0.0
Family interaction	
Relatives give assistance	14.8
Cares for adult relative	63.0
Children assist family	
No assistance	22.2
Some assistance	22.2
High level of assistance	40.8
Highest level of assistance	14.8
Owns cellphone	40.8
Health and nutrition	
Does not miss more than 5 days/month due to illness	
	33.3
Frequency of medical clinic visits	
Once a week	40.7
Once a month	40.7
Twice a year	11.1
Once a year	7.5
Frequency of meals	
Once per day	22.3
Twice per day	77.7
Eats meat, chicken or fish weekly	40.7
Livelihoods	
Raises livestock or poultry	
None	7.4
Few (< 1 point)	37.0
Medium (between 1 and 2.5 points)	48.2
High (more than 2.5 points)	7.4
Sells excess food for cash income	44.4
Has non-agricultural source of income	74.0
Transportation	
Travels once a week or more by motorbike taxi	51.9
Travels once a week or more by motorbike taxi	33.3
Travels once a month or more by bus	25.9
Travels once a month or more by car	22.2
Education	
Less than 4 years of formal education	18.5
Between 4 and 7 years of formal education	51.9
More than 7 years of formal education	29.6
Able to read and write	66.7

Source: author's fieldwork 2012.

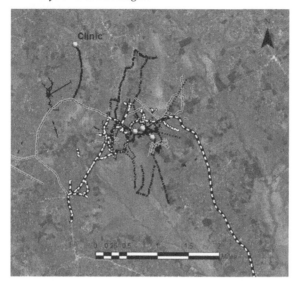

Figure 7.3 GPS tracker paths of multiple participants. (Bing satellite photos with GIS data, fieldwork by Deborah Naybor, 2012.)

they publicly dismissed a concern for making their patterns of movement known, they often left the trackers behind.

To identify patterns of normal activity, participants carried the trackers intermittently during their working hours. This data was imported into ArcMap GIS with a different color-coded layer for each woman's travel path (lines) and stopping points (triangles) and overlaid on Bing satellite photography (Figure 7.3). After collection of approximately ten to forty days of data per person over a three-month period, the researcher worked with each woman to identify points and paths using these GIS maps for each participant (Figure 7.4). Ethnographic observation assisted in the identification of each location as a site of reproductive or income-producing work, rest, social interaction or development programs (such as meetings, group work, or training). Site visits by the researcher to verify the identification of locations found that although the women could accurately identify familiar, nearby locations, more remote locations (such as the clinic or church) were often misidentified on the maps. This was most likely due to the lack of landmarks on the map in these areas or the decreased frequency of visits.

Time use and mobility of rural women

In this study, women face significant constraints that stem from lack of time to undertake daily tasks as well as inefficient ways of traveling to conduct these tasks. They spend much of their day gathering wood, finding food and preparing meals for their families. Women also spend an inordinate amount of time in reproductive labor, such as the care of children and the elderly. This research indicates that women walk to most locations—even when they must travel to the clinic. Their

Figure 7.4 Meeting for identification of GPS locations on GIS maps. (Photo by Deborah Naybor, 2012.)

mobility patterns include repetitive travel over the course of the day (such as making multiple trips from home to market) and denser patterns of movement close to home. The data showed that 40 percent of women's travel is within 800 feet of their home, indicating spatial fixity. On average, participants spent 53 percent of their time on reproductive work, including growing food for the family, household chores, childcare and other domestic duties. Almost 40 percent of their time was spent on economic endeavors like making crafts, brewing traditional beer or selling goods at markets or to neighbors. Very little time (only 5 percent) was spent on social interaction (i.e. visiting family or attending church). This percentage includes a much higher percentage of time spent in socialization by the oldest member of the group (age seventy-three), who admits she spends most of her days visiting with friends and drinking the locally brewed beer. Although they belong to a cooperative group and work on donor funded projects, only 2 percent of the women's time is spent on development meetings and training.

Time use statistics provide information that can "improve our understanding of individual and household activities especially with respect to time allocation and also improve our knowledge of the well-being of the nation" (Uganda Bureau of Statistics 2010: 42). The addition of spatial information to the study of women's mobility and time use allowed for establishment of a relationship between place and work patterns. The travel paths identified by GPS for the women who forage for food or firewood showed a distinct erratic pattern of movement as they moved through a wooded area in search of dry fuel. Most of the women reported gathering water as a time consuming daily chore. However, GPS data revealed that few women fetched the water themselves, a fact reinforced through interviews that revealed that the children are usually the water carriers in the family. The patterns of movement often reflected the inefficiency of their lives; for instance, they often made three or four trips to the same market on the same day for the buying or selling of goods, or begged for assistance from neighbors who ran the shops.

Additional spatial analysis of this data provided insight into the significance of their spatial boundaries and mobility patterns. For example, the data showed that women who work between eighty-four and ninety-nine hours a week have a higher development level and fewer health problems than those who work less than eighty-four hours a week. There are two possible explanations. First, these women lack economic opportunities, which adversely affects their income. A lower income in turn reduces their ability to seek medical treatment. But the direction of causality was not established, and it is possible that the women with poorer health also have less capability to work, which is reflected in their inability to work long hours. Also, when they spend a higher percentage of their travel distance close to home, the women increase their hours worked and are more likely to work beyond the eighty-four-hour a week time poverty threshold. This may seem paradoxical, as longer distance traveled should lead to more opportunity and therefore more hours of work. The spatial analysis using GIS reveals that, on average, the women who spend more of their "spatial budget" within 0.15 miles of home are making repetitive trips to the same places, such as the homes of relatives or the market, thus reducing the efficiency of motion. For example, taking three or four trips to the market per day, due to their physical capacity to carry a certain weight limit, wastes time, diminishes the women's energy and reduces productivity.

Conclusion

Women in underdeveloped countries are often restricted by spatial and temporal barriers, which prevent access to knowledge, education and health care, and, as a result, reduce their ability to improve their own lives. Geospatial and information technology can help these women tell their stories and map these limitations in both time and space. The potential for poor, rural women to become an integral part of the gender and development planning process is limited, in part because of the digital divide. We argue that new advances in information technology, GPS and GIS can provide critical information to enhance and advance development programs, yet the participation of women in the process is key to creating groundbreaking, successful and sustainable development.

Participatory GIS and GPS that directly involves poor women not only enriches the data but also allows women to be part of the process of developing new, innovative uses of technology by sharing their knowledge on the best way to gain in-depth understanding of the spatial relationships and restrictions caused by time poverty (Elwood 2008; Momsen 2006; Brown 2003). This participation empowers women, ensuring their representation in project design and outcome. Providing the tools of technology to women has encouraged them to believe that they are the experts and that they are capable of producing information on their own localized knowledge. Through inclusion in research design, data collection and information dissemination, the digital divide narrows and gender inequities decrease.

Women in this study reported that they increased their knowledge of their surroundings, land use and time use as a result of participating in the study. However, they felt it was particularly important that donors understand the challenges they face and that the information gathered should be shared with funding organizations and supporters in order to improve existing small-scale development programs and community planning.

As GIS becomes available to a wider audience through venues like web-based participatory GIS and applications increase, the study of time geography and accessibility holds great promise. The use of GPS technology has the capability of increasing outsider knowledge of the developing world. The potential to study the territory of bicycle hawkers in Hanoi, the movement of street kids recycling garbage in Kolkata or the range of mobility of beggars in Nairobi is now within reach of geographers using an unobtrusive 24 hours a day, seven days a week toolkit at a relatively low cost. Participatory GIS requires involvement and acceptance by the community being studied, because these approaches have the potential to both help and harm the poor. Finally, instead of reducing the role of the poor to subjects of study or mere end-users of IT, the poor must be active participants in its development and beneficiaries of the knowledge gained from the results of the investigation.

Notes

1 Participatory GIS (PGIS) employs GIS technology used by members of the public (employed by both individuals and grass-root communities) for participation in the public decision-making processes (e.g. data collection, local knowledge integration, data access, multiple representation, mapping, analysis and/or decision-making) affecting their lives. GIS has historically been considered a top-down approach, driven by data-based technology and expert knowledge-oriented techniques. Weiner *et al.* (1995) take a strong interest in making GIS a bottom-up (local knowledge) application and people-centered (participatory) approach in order to explore how GIS could integrate local and expert knowledge as well as societal issues in a spatial database meant to enhance the participatory approach in decision-making.
2 The proposed Multimedia-GIS (M-GIS) that will be available at the outcome of this project handles different forms of spatial and non-spatial information that include both qualitative and quantitative data, such as: Google Earth, GPS, spatial data, photos, narratives, videos, graphs, charts, voices and web-based information. The general public can view M-GIS data via the internet, in order to understand different aspects of female farmers' day-to-day activities, stories, livelihoods and issues. The main purpose of this data is to enable the decision-making community, professionals and community leaders to use gender-based geospatial models in development as a way of understanding gender marginalization and the digital divide in relation to access to services and information.

References

Abbott, J. (2003) "The use of GIS in informal settlement upgrading: its role and impact on the community and on local government," *Habitat International* 2: 575–93.
Abdourahman, O. I. (2010) "Time poverty: a contributor to women's poverty?" *Journal Statistique Africain* 11: 16–39.
Alagan, R. (2007) "Participatory GIS approaches to environmental impact assessment: a case study of the Appalachian corridor H transportation project," dissertation submitted to the College of Arts and Sciences at West Virginia University.

Ali, M. S. (2009) "Household accessibility analysis in developing countries using time space prism," *NED University Journal of Research* VI(1): 20–35. Available online at: www.neduet.edu.pk/NED-Journal/pdf/full/09vol1paper3f.pdf (accessed May 31, 2013).

Antonopoulos, R. and Hirway, I. (2010) *Unpaid Work and the Economy: Gender, Time Use and Poverty in Developing Countries,* New York: Palgrave Macmillan.

Asiimwe, G. (2010) "Household gender and resource relations: women in the marketing arena of income generating crops in Uganda," *Eastern Africa Social Sciences Review* XXVI(2): 1–23.

Bailey, P., Keyes, E., Parker, C., Abdullah, M., Kebede, H. and Freedman, L. (2011) "Using a GIS to model interventions to strengthen the emergency referral system for maternal and newborn health in Ethiopia," *International Journal of Gynecological Obstetrics* 115(3): 300–9.

Barbaraj, R. (2004) "Gender as a social structure: theory wrestling with activism," *Gender and Society* 18(4): 429–50.

Bardasi, E. and Wodon, Q. (2010) "Working long hours and having no choice: time poverty in Guinea," *Feminist Economics* 16(3): 45–76.

Barwell, I. (1996) "Transport and the Village: findings from African village level travel and transport surveys and related studies," World Bank Discussion Paper No. 344, Africa region series, Washington, DC: World Bank.

Blackden, M. and Wodon, Q. (2006) *Gender, Time Use, and Poverty in Sub-Saharan Africa,* Washington, DC: World Bank.

Boserup, E. (1970) *Women's Role in Economic Development,* Washington, DC: Earthscan.

Bosak, K. and Schroeder, K. (2005) "Using geographic information systems for gender and development," *Development in Practice* 15(2): 231–37.

Brown, S. (2003) "Spatial analysis of socioeconomic issues: gender and GIS in Nepal," *Mountain Research and Development* 23(4): 338–44.

Ellis, A., Manuel, C. and Blackden, M. (2006) *Gender and Economic Growth in Uganda: Unleashing the Power of Women,* Washington, DC: World Bank.

Elwood, S. (2008) "Volunteered geographic information: future research direction motivated by critical, participatory and feminist GIS," *Geo Journal* 72(3): 173–83.

Gilbert, M., Masucci, M., Homko, C. and Bove, A. (2008) "Theorizing the digital divide: an analysis of poor women's access and use of information and communication technology," *Geoforum* 39(2): 912–25.

Gill, K., Brooks, K., McDougall, J., Patel, P. and Kes, A. (2010) "Bridging the Gender Divide: How Technology can Advance Women Economically," International Center for Research on Women. Available online at: www.icrw.org/publications/bridging-gender-divide (accessed May 31, 2013).

Hanson, S. and Pratt, G. (1991) "Job search and the occupational segregation of women," *Annals of the Association of American Geographers* 81(2): 229–53.

Kabeer, N. (2005) "Gender equality and women's empowerment: a critical analysis of the third millennium development goal 1," *Gender and Development* 13(1): 13–24.

Kwan, M.-P. (1999) "Gender and individual access to urban opportunities: a study using space-time measures," *Professional Geographer* 51(2): 210–27.

Kwan, M.-P. (2002) "Feminist visualization: re-envisioning GIS as a method in feminist geographic research," *Annals of the Association of American Geographers,* 92(4): 645–61.

Kyomuhendo, G. B. and McIntosh, M. K. (2006) *Women, Work and Domestic Virtue in Uganda 1900–2003,* Athens, OH: Ohio University Press.

Lawson, D. (2008) "A gendered analysis of 'time poverty' – the importance of infrastructure," *South African Journal of Economics* 76(1): 77–88.

Memis, E. and Antonopoulos, R. (2010) "Unpaid work, poverty and unemployment: a gender perspective from South Africa," in R. Antonopoulos and I. Hirway (eds) *Unpaid*

Work and the Economy: Gender, Time Use and Poverty in Developing Countries, New York: Palgrave Macmillan, pp. 76–105.

Momsen, J. (2006) "Women, men and fieldwork: gender relations and power structures," in V. Desai and R. Potter (eds) *Doing Development Research*, Thousand Oaks, CA: SAGE Publications, pp. 43–51.

Mook, P. R. (1976) "The efficiency of women as farm managers: Kenya," *Journal of Agricultural Economics* 58(5): 831–35.

Mwaka, V. (1993) "Agricultural production and women's time budgets in Uganda," in J. Momsen and V. Kinnaird (eds) *Different Places, Different Voices: Gender and Development in Africa, Asia and Latin America*, New York: Routledge, pp. 46–51.

Noor, A., Alegana, V., Gething, P., Tatem, A. and Snow, R. (2008) "Using remotely sensed night time light as a proxy for poverty in Africa," *Population Health Metrics Journal* 6(5): 23–36.

Peet, R. and Hartwick, E. (2009) *Theories of Development: Contentions, Arguments, Alternatives*, New York: Guilford Press.

Rakai District (2010) Available online at: www.rakai.go.ug/ (accessed March 21, 2012).

Robertson, C. (1984) *Sharing the Same Bowl?: A Socioeconomic History of Women and Class in Accra, Ghana*, Bloomington, IN: Indiana University Press.

Robles, P. (2010) "Gender disparities in time allocation, time poverty, and labor allocation across employment sectors in Ethiopia," in J. S. Avabache, A. Kolev, and E. Filipiak (eds) *Gender Disparities in Africa's Labor Market*, Washington, DC: Agence Francaise de Developement and World Bank, pp. 297–332.

Roy, A. (2010) *Poverty Capital: Microfinance and the Making of Development*, New York: Routledge.

Serra, R. (2009) "Gender and occupational choices in Africa: the role of time poverty and associated risks," Pathways Out of Poverty. Available online at: www.fao-ilo.org/fileadmin/user_upload/fao_ilo/pdf/Papers/23_March/Serra_-_final.pdf (accessed May 31, 2013).

Spring, A. (2000) *Women Farmers and Commercial Ventures: Increasing Food Security in Developing Countries*, Boulder, CO: Lynne Rienner Publishers.

Uganda Bureau of Statistics (2010) "Uganda National Household Survey 2009/2010." Available online at: www.ubos.org/UNHS0910/unhs200910.pdf (accessed May 31, 2013).

UNICEF (2012) "Country Statistics—Uganda." Available online at: www.unicef.org/infobycountry/uganda_statistics.html (accessed May 31, 2013).

University of Texas, Perry-Castañeda Library Map Collection (2012) University of Texas Library, Austin. Available online at: www.lib.utexas.edu/maps/africa.html (accessed May 31, 2013).

Walker, W. and Shalini, V. (2009) "Gender and GIS: mapping the links between spatial exclusion, transport access, and the millennium development goals in Lesotho, Ethiopia, and Ghana," Resources for the Future. Available online at: http://papers.ssrn.com/sol3/papers.cfm?abstract_id=1473931 (accessed May 31, 2013).

Weiner, D., Warner, T., Harris, T. M. and Levin, R. M. (1995) "Apartheid representations in a digital landscape: GIS, remote sensing, and local knowledge in Kiepersol, South Africa," *Cartography and Geographic Information Systems* 22(1): 30–44.

Yamano, T. and Kijima, Y. (2010) "Market access, soil fertility and income in East Africa," GRIPS Discussion Papers, 10–22. Available online at: http://r-center.grips.ac.jp/gallery/docs/10-22.pdf (accessed May 31, 2013).

8 Mapping differential geographies

Women's contributions to the liberation struggle in Tanzania

Marla Jaksch

Introduction

The year 2011 marked the fiftieth anniversary of liberation from colonial rule in Tanzania and in many other African countries. Various conferences, speeches, workshops and celebrations commemorated those involved in the liberation struggle and reflected on continuing struggles for equality and autonomy in this country and world region. A closer look at the commemorative events and materials in Tanzania reveals distorted depictions of what stories are worth telling, whose experiences matter and how women contributed to this country's liberation struggle. This "pattern of forgetting" women's roles in favor of a more uniform representation of the liberation movement serves to erase women's political agency in Tanzania and across the African continent (Coombes 2003: 07).

The liberation struggle in Tanzania did not end with Tanganyikan independence in 1961, or with the creation of the Republic of Tanzania in 1964 through the union of mainland Tanzania with the islands of Zanzibar. For more than forty years, Tanzania sustained a leading role in the pan-African struggle for continent-wide independence in which women continued to play more than a supporting role. This study is based on the legacy of women's struggles to end colonial rule and the significance of Tanzania as the location for pan-African political work, safe housing for exiles and military training. Conventional accounts are not adequate for capturing the complicated and complex nature of women's contributions to the struggle and reasons for their continuous marginalization in this history.

Mohanty (1991) analyzes the silence, resistance and/or ambivalence regarding women's involvement in two-thirds of world liberation struggles and argues that Western feminists (and others) have been tempted to think of colonialism as a mostly material practice that involves political, economic and social systems of overt domination. She draws attention to forms of discursive colonialism that are less explicitly apparent but are no less significant than the scholarship and expressive culture that reproduces unequal relations of power. Her postcolonial feminist framework involves intersectionality to construct the category of "women" in "a variety of political contexts that often exist simultaneously and overlaid on top of one another" (Mohanty 1991: 65). This approach is mindful of links between women and among groups of women without falling into false generalizations, and it also acknowledges the contradictions as well as the commonalities in women's experiences.

In this chapter, I argue that creative and emergent technologies have begun to play a significant role in re-presenting the type of intersectionality proposed by Mohanty. The Virtual Freedom Trail Project (VFTP)[1] counters patterns of forgetting and silences concerning women's contributions to, and shaping of, the liberation struggle in Tanzania and provides an alternative means of telling this story. I draw from Tanzanian women activists, especially Bibi Titi Mohamed, to illustrate how their involvement in the African National Union (TANU) of Tanganyika (as Tanzania was known prior to 1964) shaped the party and was essential to the success of the liberation struggle (Figure 8.1). The VFTP centers on defining women's own identities as political actors instead of their sole identities as women, and on how each of these identities transforms the other.

New media methods used in the VFTP are adapted to better understand how these activists conceptualize themselves not only socially and temporally, but also by highlighting the importance of space and place to their development as political agents. Many of the women involved in the establishment of TANU are Muslim and semi-literate and are therefore highly restricted in some of their movements. Yet, these women were able to negotiate these restrictions and, in a very short period of time, recruit thousands of members. Mapping women's spaces and spheres of influence challenges the patterns of forgetting and makes visible a more comprehensive trail to freedom through the living rooms, kitchens, market places and alleyways to stadiums and political halls that served as political education and recruitment sites. These spaces contain stories that extend beyond the passive setting for historical action to profoundly shape the type of political organization that TANU became. This attention to space, along with intersectional approaches to re-telling women's stories that challenge normative depictions of African women, allows us to map and make sense of more complex and contradictory textual, oral and visual representations of women in the liberation movement. In doing so, I contribute to a wider debate regarding feminism, nationalism and transnational struggles.

The VFTP uses emergent technologies, such as feminist Geographic Information Systems (GIS), to capture and map a virtual trail with sites of resistance as complex living spaces. These virtual maps also focus on the intersectionality and transnational aspects of women's experiences and locations to generate alternative forms of women's knowledge production and function as a form of "differential mapping" or "thick mapping" (Drucker *et al.* 2012). Mapping differential geographies has the potential to contribute significantly to a newly emerging transdisciplinary literature that combines participatory-GIS (PGIS) with historic conservation, cultural heritage and commemoration. These techniques are applied here in economically poor countries where gender inequality is manifest in land use/rights and the digital divide (Nieves 2009). Kwan (2002) argues that GIS can be re-envisioned and used as a method that enriches feminist research. Several feminist geographers have proposed the use of feminist visualization to open up new possibilities for representation (Kwan 2002; McLafferty 2002). In addition, feminist geography is particularly attentive to the construction of gendered bodies

Figure 8.1 Screenshot of the "mapping" of Bibi Titi Mohamed. (Photo by Marla Kaksch and Angel David Nieves, 2012.)

and identities as well as the extent to which GIS can represent these social dynamics. Moreover, as envisioned here, the VFTP functions as a form of feminist commemoration, living museum and digital archive. Accordingly, arts and cultural institutions can serve as complex sites of knowledge production in the postcolonial context. In constructing epistemologies of representation, the arts have the potential to validate certain knowledge claims while discounting and erasing others.

This chapter explores the reasons behind the omission of women's contributions to the liberation struggle by examining the politics of gender, memory and representation in arts and culture and by providing examples of discursive colonialisms in nationalist and Western feminist historiographies. Projects like the VFTP offer alternative ways of telling liberation history and illustrate how they might challenge some of the silencing tendencies prevalent in other methods. Interpretive historical practices, together with thick mapping of disparate parts of the liberation struggle (such as visual and textual materials, songs, maps and stories), resist glorified representations of women as freedom fighters (Lyons 2004) or as hapless victims (Mama 2001). This analysis focuses on the life and work of Bibi Titi Mohmed and the complex nature of women's stories and representations in ways that have an impact on not only the telling of history, but also women's ability to sustain an authentic political life and agency (Mohanty 1991).

The gendering of liberation struggles in Tanzania

The fiftieth anniversary of liberation from colonial rule marked a new trend of African heritage programs with the re-dedication of museums and the conception of cultural commemorations and tourism projects. The United Nations Educational, Scientific, and Cultural Organization (UNESCO) announced a funded project, "Roads to Independence: African Liberation Heritage Programme," in December 2011, that would be headquartered in Tanzania. Early UNESCO documents about the proposed heritage center acknowledged a number of challenges to the proposed project, noting that:

> *many women in the liberation movements have not had their stories documented and this would be an opportunity to make visible the critical role women played in the struggles for independence in Africa* (emphasis in original).
>
> (UNESCO 2005: 7)

In a *Daily News* article referring to the multi-national heritage project, the Tanzanian prime minister echoes the call for swift action and collaboration in documenting and preserving the rapidly disappearing history of African liberation in Tanzania. In reference to the proposed UNESCO project, Prime Minister Mizengo Pinda observed that:

> in a world still dominated by men, society tends to ignore the roles played and contributions by women. The liberation struggle in our continent involved both men and women. We are all aware of contributions made by great women

such as Mama Winnie Mandel and Mama Josina Machel who died during the struggle. This programme must therefore recognise and document those women's stories and heroic deeds alongside those of men.

(Kiisheweko 2011)

The article suggests the recognition of women's stories and action in the struggle, but, like Pinda, neglects to mention a single Tanzanian woman by name, as if the important contributions made by women happened in places outside of Tanzania or by women who were not Tanzanian. Susan Geiger, a Kenyan journalist visiting Dar es Salaam in the mid 1980s, asked her, "Why would you want to know about women in Tanzania's nationalist movement? Surely this is boring compared to Mau Mau and our fight against white settlers!" (1997: 8).

According to many scholars, social and collective memories are partial and selective (at best) and, as such, they are constructed representations of the past as much as they might be an accurate depiction of truth or fact (Bold *et al.* 2002; Coombes 2003; Miller 2009, 2011). In this way, memories of national struggle, achievement and trauma are often used to serve certain political purposes and for that reason are hotly contested (Schroeder 2012). In considering public memory in post-apartheid Africa, Coombes suggests that the vital role of women has been "sorely neglected in favour of a more monolithic representation of the liberation movement" (2003: 107).

Despite the erasure of women's participation and location in liberation struggles across the continent, many women were active in long-term national liberation efforts (Horn 2012). This involvement is evident in speeches, songs, artwork, images and testimonies as well as through the attention to women's roles in liberation politics from male heads of liberation groups and presidents of newly independent countries (Horn 2012; Hunt 1989; Jaksch 2013). For example, during a 1960 speech to the Conference of Women of Africa and of African Descent in Ghana, Kwame Nkrumah, former leader of pan-African liberation and first democratically-elected president of Ghana, asked, "What part can the women of Africa and the women of African descent play in the struggle for African emancipation?" (*Evening News*, Ghana 1960). He went to on to mention the many ways women had and could continue to support the cause of liberation.

Samora Machel, the military leader of the Mozambican liberation group Front for the Liberation of Mozambique (FRELIMO) and the first freely-elected president of Mozambique in 1975, was very outspoken about the role of women in the liberation struggle and beyond. Barry Munslow quotes Machel as saying that:

Because most of us were educated by the exploiters we have come to think that manhood has something to do with the domination of our women. . . .It is also very important that women see the necessity of political power because it is the only way to liberation of the people.

(Munslow 1985: 17)

Nyerere also addressed issues of gender inequality regularly in speeches and in important founding documents, including the Arusha Declaration.

Despite this attention to women's roles in the liberation movement in Africa, the writings of Nkrumah largely document the political thought driving the independence struggle in Ghana. However, the political process and tactics of women who were central to liberation in this country is largely forgotten. The work of women is turned into private and ordinary acts of less significance. Women are represented flatly as blind participants rather than as leaders in ideas and strategies. Even the political thought of women who have held formal leadership positions in liberation struggles is relegated to the margins. For example, the work of political leaders such as Field Marshal Muthoni in Kenya, Thenjiwe Mtintso in South Africa and Bibi Titi Mohammed in Tanzania has yet to be completely documented and archived (Horn 2012). Feminist scholars Alexander and Mohanty (2010), McFadden (2005) and Mama (2001) agree that to fully understand the meaning of the struggle, it is vital to know how women's knowledge shapes the movements in which they hold active leadership roles. Mama (2001) argues that feminism originates in Africa and has provided clear examples of African feminist epistemologies. Furthermore, Miller (2011) work on women and commemorative practices in post-apartheid South Africa, asks how historical truth is possible without an account of women's experiences.

Historian Geiger (1997) suggests at least two reasons for women's exclusion from the history of the liberation struggle in Tanzania. First, she explains that the marginalization of African women in history tends to reflect an all too familiar pattern of androcentric bias in recorded history. This process of erasure is accomplished through successive accounts in primary (produced by colonial officials/travelers) and secondary (produced by Western and African scholars) records that echo the silence of the previous one. Geiger also proposes that the erasure of women has been more thorough in accounts of Tanzanian nationalism written in English than in those texts written by Tanzanian scholars and activists writing in Swahili. Given the dominance of English publications, these omissions become unquestioned historical facts that have remained stubbornly fixed for decades.

Geiger (1997) suggests that traditional frameworks for understanding nationalist, colonial and "traditional" accounts of history in the Tanzanian context privilege *ujamaa*, a founding concept developed by Julius Nyerere (1967), which emphasized cultural unity over ethnic, geographic, religious, linguistic, economic and other differences. *Ujamaa* is an economic and social development model that influenced all areas of life including scholarship/education, culture and the arts.

Moreover, Berger (2003) highlights a masculinist and patriarchal tendency in Western historiographies that focuses almost exclusively on the experiences of men. When they appear, women are depicted as victims or marginal figures with no subjectivity or agency (McClintock 1993). Geiger (1997) also challenges dominant Western feminist frameworks about women and nationalism by providing a rich contextualization of Bibi Titi Mohamed's leadership in TANU. She raises many questions and contradictions and writes that the dominant narrative does nothing, however, to explain:

> *how* Bibi Titi Mohammed, a 30-year old lead singer in the popular ngoma (musical) group, "Bomba," became an actual rather than fictive TANU leader, and *why* thousands of women became nationalist activists as a result of her political acumen and enthusiastic work.
>
> (Geiger 1997: 11)

Through interviews, oral histories and testimonies of women involved in the liberation struggle, it is evident that women remember this time very differently than men. For example, Geiger notes that the women she spoke to offered a:

> very different picture in their construction of the present and the reconstructions of their past political involvement. Their narratives show that in the 1950s, "ordinary" (illiterate frequently self-identified Swahili, Muslim) women created and performed Tanganykian nationalism "culture of politics."
>
> (Geiger 1997: 15)

These stories map the trajectory of the movement by reframing liberation as a living process rather than a thing that happened. Another body of writing by women engaged in colonialism further erases African women, who "disappear into the blind spot of Western gender and racial ideologies" (Geiger 1997: 10). Involved in some way with the colonial enterprise, wives of colonial officers, missionaries and settlers often wrote about what they witnessed with concern and fear. But much of this work is riddled with particular Western gender ideologies interconnected with racist imperialism and the construction of the colonized woman as "the other's other" or, as Mohanty would say, "through acts of discursive colonialism" (1991: 65). Spivak (1999) expresses her concern for the processes whereby postcolonial studies re-inscribe, co-opt and repeat neocolonial imperatives of political domination, economic exploitation and cultural erasure. This essay raises questions about whether or not feminist and postcolonial critics are unknowingly complicit in doing the work of imperialism.

Finally, a significant contribution to the literature of women freedom fighters is White's (2007) history of the role of "guerrilla girls" in anti-colonial struggles. In many cases, women trained in anti-colonial armies were not included for reasons of equality or feminist consciousness but rather because of necessity. She explains:

> In contrast to Fanon's claims about revolutionary violence as a cleansing force, war is a dirty business and a gendered business. Rather than serving as a transformative, humanistic force, in many contexts violence functions as a degenerative force. The trauma and humiliation caused by debilitating violent acts left many women soldiers serving in anticolonial forces feeling unworthy of any recognition, much less mutual recognition.
>
> (White 2007: 78–9)

White (2007) argues that anti-colonial warfare works to re-establish dignity for colonized men because it restores traditional norms of masculinity, but this same warfare violates gender norms when women participate. Because of this violation, women frequently lie about participation in order to avoid censorship or exclusion from society. As a result, female revolutionary makers of history have been pressured to disappear from the narrative of history. The following section examines how the arts and cultural institutions in Africa have served as complex sites of knowledge production in the colonial and postcolonial context. In constructing epistemologies of representation, the arts have the potential to validate certain knowledge claims, while discounting and erasing others.

Memory, gender and the politics of representation

Historically, cultural institutions like museums, art galleries, archives and art academies were established either by the colonial state or in the context of post-colonial nation building in most African countries (Askew 2002). Consequently, the cultural field has often been shaped according to a variety of tensions that reflect national aesthetics and/or thematic concepts and guidelines established by the West. In Tanganyika, both the German and the British colonial governments were actively involved in destroying existing theatre, dancing and other cultural practices, institutions and activities, because they were deemed to be "uncivilized activities" (Lihamba 2004: 236).

Although they occupied Tanzania, the British were less committed to transforming it into a formal colony like Kenya (Brennan 2012). Nonetheless, the British were involved in establishing a few institutions they considered worthy, including the Dar es Salaam National Museum, dedicated to King George V. In 2011, the museum was re-dedicated as "The House of Culture" to celebrate the history of Tanzania; it includes a variety of exhibits such as the bones of *Zinjanthopus boisei* (the findings of the Leakeys in Olduvai), Shirazi state items, ethnographic collections from various ethnic groups, Tanzanian history from early societies to the present day and a small, loosely organized collection of materials related to German and British rule (www.houseofculture.or.tz; accessed October 28, 2013).

Starting in the 1920s, the British occupation resulted in the forceful introduction of colonial theater. Between the years of 1945 and 1952, colonial theater was aggressively taught and presented not only by the British government, but also through missionaries. Although it had no roots in Africa, colonial theater was portrayed as universal (Mollel 1985) and presented a challenge to the widespread interpretation that traditional performance was related to things that the Church sought to deny to African people, from "indigenous religion, to sexuality and alcohol" (Plastow 1996: 45). Further, scholars such as Bakari and Materego (2008), Kerr (1995) and Nsekela (1984) argue that the banning of such subjects was no accident. Nsekela arges that the colonial education often provided by missionaries was used as a tool to encourage people to accept "human inequality and domination of the weak by the strong" (1984: 58) as a fundamental element of being civilized. Although they were able to ban many traditional African performing arts, the British generally tolerated *beni ngoma*, the mostly male costumed dancers and musicians who performed skits.

Arts and culture also served as a site of subversion and anti-colonial organization. Many members of *beni* associations were a part of the nationalist movement that gave rise to the Tanganyikan African Liberation National Union (TANU), the party that fought for independence in 1961. While many of these *ngoma* organizations were just for men, there were many women-only and gender-mixed groups as well. Religious and other *ngoma* groups, as well as other forms of art, also played a large role in liberation politics. Through *ngoma*, Bibi Titi Mohamed became the leader of the women's section of TANU and later the sole woman in post-independence government as part of President Julius Nyerere's inner circle. Bibi Titi has been quoted as saying that being a part of a singing group led her to politics and being successful in recruiting members to TANU. She states:

> Nobody knew anything about independence or what it was. People were afraid….But we approached them slowly and tactfully, explaining what it meant, and where we were going, and the meaning of what we were doing. And they accepted. But my position was helped by these people of the organizations.
>
> (Bibi Titi Mohamed in Geiger 1997: 50–1)

Another example of the role of the expressive culture shaping politics is the *khanga*, translated from the Swahili verb *ku-kanga*—to wrap or close—a colorful piece of printed cotton fabric worn by women that includes a border along all four sides with a proverb printed on one of the long sides (Figure 8.2).

Elinami Veraeli Swai argues that *khangas* have functioned as art, fashion and "important inventories of knowledge as well as archives of social and political commentary" (2010: 81). In contrast to many written accounts of women's fashion

Figure 8.2 Khanga with political message *Umoja wa Wanawake wa Tanzania* (Union of Women in Tanzania). (Photo by Warcheerah Kilima, 2011.)

in Africa as being static and lacking in creative energy or fashion consciousness, Swai draws upon the work of Jean Allman (2004), who notes that there is power associated with fashion. African women have used fashion to represent themselves, their interests and their ambitions. Allman also argues that all forms and manner of dress are an "incisive political language capable of unifying, differentiating, challenging, contesting and dominating" (2004: 1).

In addition to Allman's assertions that clothing has more importance than simply making a fashion statement, changing lives and shaping behavior, Swai claims that women use it as a source of knowledge systems in Africa—a literary genre, where "knowledge is produced, used, and disseminated mostly by women" (2010: 82). The utilization of *khangas* legitimizes women's knowledge systems, whereby creators, sellers and wearers of this art form have the ability to shape people's everyday thinking and behavior. In this case, *khangas* reveal the ways women have been speaking their stories and speaking back all along. As a form of art, they serve as a device to "breakdown barriers and penetrate areas that women could not reach with spoken words" (Swai 2010: 85), and since they are worn in public, they have the ability to be shared with large audiences. *Khangas* are also sources of history, and since they are created by ordinary women, they manifest as diverse commentaries on social, economic and political issues; they have the ability to disclose women's thoughts and ideas.

Swai (2010) also argues that *khangas* are a unique art form that is deeply integrated into the development of a national identity. During the 1960s, many *khanga* themes were explicitly about *uhuru* (independence), postcolonial celebration and access to resources (Figure 8.2). In times of liberation, they expressed desires such as *uhuru ni kwa wote* (independence is for all) and *wanawake pia wantaka masoma* (women also need to be educated). By the 1970s, *khanga* themes began to reflect grievances against the post-colonial state, with examples such as *wanawake walipigania uhuru pia* (women also fought for independence) and *mbona wanawake wamesahaulika* (why have women been forgotten?).

Although many historians have not read *khangas* as a significant art form connected to liberation and the shaping of postcolonial identity, one can see that the arts in general were a part of Nyerere's overall plan to unify Tanganyikans. The significance of art in developing a national identity in Tanzania can be seen through the establishment of the Ministry of Culture, which was the earliest post-independence initiative Nyerere created to fight against cultural imperialism. The Julius Nyerere Cultural Centre, *Nyumba ya Sanaa* (The House of Art), was also established to provide a location for artists to promote their work and offer workshops. The building itself was home to many original works of art by Tanzanian artists. However, some Tanzanian scholars (Mlama 1985) suggest that directing cultural policy toward a capitalist model of selling reveals a lack of interest in moving toward socialist construction. Therefore, the weaknesses of art institutions in the past show a disconnection between the liberation and post-independence rhetoric related through specific arts and culture-based policies and the lack of leadership. Mlama argues that lack of a defined socialist cultural policy has led to a "culture from above" approach that attempts to impose culture from above rather than in the name of developing a "national" culture (1985: 17); the cultural needs of the people at the grass-roots level go unattended. In many ways, this assertion is significant. Moreover, the instability and lack of development of many arts and cultural institutions are due to a lack of capital (rather than a lack of leadership) and a fear of ethnic divisions that have produced weak, top-down approaches, unconnected to the vast majority of the population.

Although many African artists and activists have aligned themselves with state-initiated cultural politics, others have criticized and distanced themselves. Examples include community archives and community art centers in apartheid South Africa, or initiatives like "Laboratoire Agit-Art" in post-independence Senegal—where new types of spaces and initiatives were created outside of the previous models and frameworks. They developed apart from municipal and/or state-affiliated institutions as well as separate from commercial (art) markets in order to create alternative models and platforms for negotiating and reflecting upon art and history.

Some of these examples aim to establish self-organized, non-hegemonic and experiential fields and orders of knowledge, while others deliberately question institutions developed by the postcolonial nation-state and attempt to fill in where public institutions are undermined. In many cases, scholars, cultural practitioners, curators, artists and activists join together in order to collaborate in these spaces. Through new forms of South–South cooperation, innovative media, information and communication technologies, transnational networks and knowledge are being developed. These highly creative and collaborative ways of (re)constructing history provide a background to developing the Virtual Freedom Trial Project.

Societies emerging from conflict, oppression and violence face numerous challenges. A variety of strategies exist for "healing" the wounds of the violent past and for coping with the future (such as national reconciliation and peace building). These efforts can be promoted through effective use of innovative commemorative and heritage projects for empowerment, while providing redress for marginalized communities. Well-established forms of heritage pageantry that include national memorials and commemoration celebrations are often more about national performance than social justice. Through alternative heritage projects such as the VFTP, important histories of the "other" that were formerly overlooked, silenced and diminished may now be explored (Coombes 2003; Miller 2011; Nieves 2009).

In the Tanzanian context, tension exists between the rich political arts traditions, cultural practices and arts and post-independence cultural policies across the continent. Women's heavy involvement in the liberation struggle was not evident in a pervasive use of images of women as liberation figures (which was common in South Africa and Zimbabwe, for example). Yet the arts still played a critical role in terms of the space it provided for women's political agency. In her attempt to understand the role of so many women involved in the nationalist movement in Tanzania, Susan Geiger asked, "[H]ow do we explain the fact that whatever else its characteristics, Tanganyikan nationalism has remained remarkably devoid of the symbolic objectification of women?" (1997: 13). And yet, the majority of the cultural work in Tanganyika was undertaken by women.

In a 2012 *Pambazuka News* article about whether or not women in Africa are now occupying new movements, the author pinpoints the sense of shock uttered by many people that women play such a large and visible role in the many recent acts of mass civil resistance. According to Hakima Abbas, even though women have been widely involved in past liberation struggles:

> [W]hat is consistently disputed is the participation of women in acts of civil resistance and while the question of role is posed, an important question of power remains. In the age of Twitter, Facebook and rapid diffusion of images around the world, *the participation of women on the front lines of protest and in acts of mass civil resistance can no longer be disputed* (emphasis mine).
> (Abbas 2012)

The following section explores the possibilities of new and emergent technologies in relation to innovative trends in commemoration and heritage projects that include the creation of the VFTP as a way of countering the erasure and silencing of women's contributions to liberation struggles in Tanzania.

Mapping differential geographies of African women's contribution to liberation struggles in Tanzania

The Virtual Freedom Trail Project lies at the intersection of efforts to rethink African liberation history and heritage and debates surrounding how these measures should be undertaken and who should directly benefit from them. Differential mapping highlights the complex relationships that trace women's contributions to the struggles for liberation in Tanzania and South Africa. This project examines archival, visual and textual representations of women political activists in Tanzania, including female anti-apartheid and liberation activists living in exile there (both during and after the struggle against apartheid and colonialism). Through the use of thick and differential mapping, I examine women's participation in the struggle for democracy and how it is represented, remembered and often forgotten in contemporary Tanzania's history, visual culture and commemorative/heritage sites. In both countries, ongoing debates about national transformation and women's continued inequality largely forget the complex history of political identities and recognition for women.

Scholars interested in documenting women in a global context have been highly motivated by international feminist movements that continue to maintain a political component to much of the research and writing in African women's history (Mulligan-Hansel 1995; Perter and Tenga 1996). Sheldon suggests that this work was possible by showing that women "were present and active despite their omission" from traditional historiography (2010: 192). However, researchers were hampered by the paucity of published and archival materials related to women in Africa. Therefore, they developed new methods to retrieve historical information, including interdisciplinary approaches such as geography and linguistics and analyzing women's expressive culture (tattoos, textiles, arts and songs). Geiger's (1996) life histories and oral testimonies have been central to capturing aspects of women's lives and to filling the void in collecting information about ordinary women's lives. However, in recent decades, the development of digital-based projects, which include digitizing existing archives and using new technologies to enhance or expand types of oral, video/diary histories and testimonies, has increased, but few have been made available to the public. Drawing on the prototype for Soweto'76 (www.soweto76archive.org/3d/video/; accessed October 28, 2013) and the Soweto Historical GIS Mapping Project, the VFTP expands on this dynamic work and includes community member and student contributions through multi-authored publishing in a dynamic platform (Figure 8.3).

Bodenhamer (2010) argues that major developments in GIS in the last twenty years have been so profound that their impact on many areas (e.g. commerce, academia, government and administration) has been equated to the advent of key technologies like the microscope or the printing press. And while GIS has been most utilized in the sciences and social sciences, there is now an effort to adapt this technology to enhance research in history, cultural studies and the arts; a fusion that is now typically referred to as the spatial humanities.

According to Bodenhamer (2010), as inherently spatial beings, people occupy a physical world that necessitates using spatial concepts such as distance and direction in order to make their way through it. He goes on to argue that while our typical associations with space are often subconscious and mundane, as researchers, we are drawn to the ways in which space offers us a way to understand how we fundamentally order our world. For example, through contemporary notions of space, "We acknowledge how past, present, and future exist in constant

Figure 8.3 Soweto Historical GIS Project team with local informants using ArcGIS on iPad. (Photo by Angel David Nieves, 2012.)

tension with other representations from different places, at different times, and even at the same time" (Bodenhamer 2010: 14).

There is also the recognition that past, present and future conceptions of the world are in simultaneous competition in real and imagined spaces and are always contingent. Bodenhamer contends that the meaning of space is constructed through power relations. Feminist geographers have for some time now pointed to the ways in which space is identified through gendered tropes (Mother Nature) and in narratives about colonialism, nationalism and liberation. Tanganyika is often constructed as a feminized space due to the forms of colonialism it endured. In particular, it is "Tanganyika's Mandate, and later Trusteeship, status" (Geiger 1997: 8) that marks it as something other than a "real" colony and suggests a reason for its exploitation and marginalization from histories of liberation struggles.

South Africans seeking refuge in Tanzania landed in a specific historical context that included Tanzania's struggle as a newly independent country, the Cold War and an era of violence on the continent. While there had been waves of South African exiles secretly making their way to Tanzania for decades, it was not until the student uprisings in 1976 that a massive influx of exiles, mostly youths, including many women and girls, arrived in Tanzania. In her edited collection of stories of exile from South Africa, Bernstein (1994) writes that many of those that fled were students, who often left the country illegally and on foot. With the growing number of exiles, it was decided that schools and training centers needed to be established in Tanzania. This country was the site of more than a dozen different types of shelters for war refugees, safe houses, conference locations, diplomatic spaces and military training bases that reinforced the distinctive geography of the liberation struggle. According to Schroder, a lot of attention has been paid to the significance of:

> "external" contributions by the international solidarity movement in comparison to the "internal" struggle being waged by combatants in South Africa itself. In this regard, Tanzania constituted neither an "external" nor an "internal" force, but instead occupied a third, interstitial space that remains relatively underexplored in histories of the period.
>
> (Schroder 2012: 11)

However, information about the many youths that fled South Africa to receive educational and vocational training (as well as military training) in unique, largely self-reliant model communities, remains largely undocumented.

Scholars have widely accepted the notion that a lack of extant archival sources in Tanzania is due to the political restrictions placed on the African National Congress (ANC) in South Africa. Interestingly, ANC materials were almost always destroyed to prevent incriminating evidence from falling into the hands of the apartheid state. In Tanzania, however, many of the documents and papers related to the struggle against apartheid are extant and untouched due to a lack of resources for promoting their conservation. As a result, much of this history is simply unknown to researchers. A unique opportunity exists to begin the process of cataloguing these extant material artifacts, documents and holdings from these schools and settlements.

In order to achieve a "network of nested multi-modal narratives," the use of multimedia is essential. The project is realized in a multimedia format because of the need for real-time witnesses and the urgency of claims made by women and their families, currently in Tanzania and elsewhere, who have been ignored by the ANC government for nearly two decades since the end of apartheid rule. ANC exiles, primarily women, have often complained about the fact that any kind of democratic accountability was subverted by the ANC in exile (Bernstein 1994). Forced family planning, sexual abuse, rape, abandonment and extreme poverty are among the tragic outcomes of the Tanzanian Diaspora (Morrow *et al.* 2004).

The case of Bibi Titi Mohamed

Bibi Titi Mohamed (1925–2000) is an important and complicated figure in Tanzanian liberation history. After Geiger published her work on TANU women in 1997, little attention was paid to Mohamed (or the thousands of women who worked with her) until liberation commemorations were held across Tanzania in 2011 and 2012. This oversight is partly due to the notion that Mohamed's story complicates dominant nationalist history in Tanzania. Specifically, life history narratives of women activists in TANU disrupt the view that liberation happened in progressive stages to develop a nationalist consciousness that borrowed heavily from Western forms and ideals (Geiger 1996).

On July 8, 1955 Mohamed held her first political meeting of the Women's Section of TANU, in which more than 400 women were said to have joined the organization. By October of the same year, Mohamed enrolled 5,000 women members. The TANU organizing secretary, Oscar Kambona, wrote to London that although only semi-literate, Mohamed was "inspiring a revolution in the role of women in African society" (Kambona as quoted in Geiger 1997: 11). Geiger argued that the understandings she assembled from her interviews with TANU women did not easily fit with existing explanations of nationalism—in part because the latter denied the contributions of ordinary, if semi-literate, Swahili women. Understanding the success of Mohamed and TANU required admission that these women profoundly shaped TANU values and thus made it a relevant and viable political movement for both women and men. This admission is counter to the dominant narrative depicting a few men as "proto-nationalists whose anti-colonial actions set the stage" for later Western-oriented political work of nationalists in Tanzania (Geiger 1996: 465). Women's stories confirm that their lived experience (which reflected trans-tribal ties and affiliations) did not merely respond to TANU's political rhetoric; these women "shaped, informed, and spread a nationalist consciousness for which TANU was the vehicle" (Geiger 1996: 465).

Mohamed and her fellow activists' political work did not end with the struggle for independence, but was instrumental in the continuation of women's political culture of nationalism in post-independence years through efforts like the spearheading of the formation of the All Africa Women's Conference in Dar es Salaam in June 1962. She held a short-lived position in parliament and within TANU leadership, until she resigned due to conflicts over ownership of property as outlined in the Arusha Declaration. But her work also came into conflict with male TANU leadership, and, in November of 1962, President Nyerere declared that all women's organizations should be dissolved and merged into one national women's organization, *Umoja wa Wanawake wa Tanzania* (UWT). She led UWT until she resigned in June 1967 as a result of conflict within the organization. In 1969, Mohamed was arrested and detained on charges of treason. She and four others were tried, found guilty and sentenced to life in prison for attempting to overthrow the government and kill Nyerere. She served two years and two months before she was released. President Nyerere pardoned her in 1972.

One of the many acts undertaken by the new government to create unity and to celebrate liberation after independence was to rename many of the major roads in the city of Dar es Salaam in honor of significant heroes of African liberation history. Mohamed was recognized as a TANU leader and a road was named in her honor (Figure 8.4). This recognition of Mohamed is interesting given the reluctance to acknowledge individual women as leaders in the liberation struggle and given the few memorials that exist in Tanzania commemorating liberation history. After Mohamed was imprisoned, the road was renamed *Umoja wa Wanawake wa Tanzania* (Union of Women of Tanzania) but was eventually changed back again in the mid 1980s during the thirty-year independence celebrations.

In order to reverse the omission of women and their contributions to the liberation struggle, the VFTP incorporates visual ethnographies, audio diaries/life histories,

Figure 8.4 Road named after Bibi Titi Mohammed in Dar es Salaam, Tanzania. (Photo by Marla Jaksch, 2009.)

digital mapping and archival evidence with resistance sites as complex living places (Figure 8.5). Much of this work has begun but has yet to be analyzed and differentially mapped to date; however, I will begin the process of using ArcGIS to trace specific locations around Dar es Salaam (with women who are former TANU) with the goal of mapping the spaces of memory, community, culture and religion with those locations that no longer physically exist.

The use of thick or deep mapping demands that we focus on experiential navigation, epistemologies of representation and the rhetoric of visualization (Drucker *et al.* 2012). Essentially, maps are visualizations or representations of a group of relations that present a state of knowledge. Digital maps can function as interactive sites for creating, representing and navigating layers of spatial data. Drucker *et al.* (2012) argue that thick mapping allows us to move away

Figure 8.5 Virtual Freedom Trail Project (VFTP) map. (Designed by Gregory Lord, Digital Humanities Initiative, Hamilton College; photos by Marla Jaksch and Angel David Nieves, 2012.)

from a map as a static representation of a physical reality. Within this dynamic environment:

> [N]ew data sets can be discovered, and perhaps most importantly, missing voices can be returned to specific locations through "writerly" projects of memory that the participatory architecture of Web 2.0 applications have made possible. Thick mapping thus enables us an unbounded multiplicity of participatory modes of storytelling and counter-mapping in which users create and delve into cumulative layers of site-specific meaning.
>
> (Drucker *et al.* 2012: 47)

The larger project includes a collaboration of small working teams, consisting of researchers, scholars, community members and students. Each team has specific work on developing a "network of nested multi-modal narratives" in relation to a through media-rich, interactive "map" of the accompanying historic site (Figures 8.1 and 8.5). These media-rich site maps feature details about each location within Tanzania and South Africa and are currently being mapped and linked (or nested) to testimonies, inventories, bibliographies, scholarship, historic images/3D models, geospatial data (including GIS), texts and audio. This mapping is not intended to result in merely describing the topographic and spatial features of resistance, as manifested in several sites in Tanzania and South Africa. Rather, the project requires the utilization of creative and innovative methodological approaches—especially those of the spatial humanities that will allow for a better, more complex understanding of what is commonly referred to as heritage resulting from state oppression and violence in post-conflict societies.

Conclusion

Recent events have brought momentous causes for celebration and dramatic shifts in women's empowerment on the continent of Africa. In 2010, the African Union declared the beginning of the "African Women's Decade." In 2011, the visionary Kenyan leader, Wangaari Mathaai, the first female African Nobel Peace Prize winner, passed away but one week later two Liberian women, Leymah Gbowee and President Ellen Johnson-Sirleaf, joined her in history as Nobel laureates representing different but compelling models of activist leadership. Malawi saw the emergence of the second African female president (the first being Johnson-Sirleaf), Joyce Banda, and in Uganda, the outspoken Proscovia Allengot Oromait became the youngest Member of Parliament (MP) at the age of nineteen in 2012. These historic achievements seem impossible to understand without knowing the rich history that led to this rise in leadership and activism.

Forgetting women's contributions to liberation not only erases their political agency but also impacts possibilities for acknowledging the achievements of women in the future. Culture is a common source of wealth and a tool for advancing women's equality that is achieved through policy and gender quotas as well as by allowing people to have access to what constitutes cultural identity, history and heritage.

Further, the Morogoro community (among others) has expressed an interest in finding alternatives to non-sustainable livelihoods (environmental degradation) by investing in heritage as a source of cultural identity (Figure 8.5). In this way, cultural property can provide opportunities for tourism and development. Often, what is defined and financially supported as culture by outside experts reinforces colonial/neocolonial ideas and also takes control of, and benefits from these cultural development projects. In fact, part of the pattern of forgetting women may have to do with Tanzania's shift to a neo-liberal economy. Schroeder (2012) suggests that patterns of forgetting may in fact be related to a process of organized forgetting, or forgetting from above, that is symptomatic of the process of neoliberal reform. Pitcher defines "organized forgetting" as a conscious process of "dissociation from the past, engaged in for the purpose of constructing a new ideology, creating new institutions and organizing new networks to confront the present" (2006: 89). The rationale behind this is that erasures and distortions support neoliberalizing regimes by suppressing populist sentiments and communitarian values connected with earlier, socialist nation-building projects (Schroeder 2012: 4).

In this transnational feminist project, I seek to make visible the complex representations of girls and women that might translate into substantive changes to women's rights and improve material conditions (Ndziku 1994). This may be possible through the adoption of new methods and epistemologies for commemorating and representing women's contributions throughout history. GIS-facilitated understandings of society and culture that embrace multiple voices, views and memories of our past, while available to be viewed on different scales, have vast potentials. The use of new media offers up a host of complicated questions and possibilities. Often, one hears about technology dumping or the "transfer" of knowledge, but this project conceives of aspects of technology in a slightly different way. Kwan (2002) argues that GIS can be re-envisioned and used as a method that enriches feminist research. Feminist geography has been particularly attentive to the construction of gendered bodies and identities and the extent to which GIS can represent this. Work such as Kwans's suggests that there are many representational possibilities of GIS yet to be explored. Through different methods that use qualitative and quantitative data in real and conceptual space (such as feminist visualization in GIS), this project has the potential to address the ambivalence regarding women in Tanzania, whether it be in better representing them in the past liberation struggle or in the present, as political agents. Such a step might make visible barriers that continue to block the freedom trail for women by providing an alternative view of history and culture through a dynamic representation of memory and place that is visual and experiential.

Note

1 The Virtual Freedom Trail Project began in 2009 with co-researcher Angel David Nieves, Associate Professor and Co-Director of the Digital Humanities Initiative (DHi) at Hamilton College, NY, USA.

References

Abbas, H. (2012) "Are women occupying new movements?," *Pambazuka News*. Available online at: http://allafrica.com/stories/201206290355.html?viewall=1 (accessed June 4, 2012).

Alexander, M. J. and Mohanty, C. T. (2010) "Cartographies of knowledge and power: transnational feminism as radical praxis," in A. L. Swarr and R. Nagar (eds) *Critical Transnational Feminist Praxis*, Albany, NY: SUNY Press, pp. 23–45.

Allman, J. (2004) *Fashioning Africa: Power and the Politics of Dress*, Bloomington, IN: Indiana University Press.

Askew, K. M. (2002) *Performing the Nation: Swahili Music and Cultural Politics in Tanzania*, Chicago, IL: University of Chicago Press.

Bakari, J. A. and Materego, G. R. (2008) *Sanaa kwa Maendeleo: stadi, mbinu na mazoezi*, Moshi: Viva Productions.

Berger, I. (2003) "African Women's History: Themes and Perspectives," *Journal of Colonialism and Colonial History* 4(1).

Bernstein, H. (1994) *The Rift: The Exile Experience of South Africans*, Ann Arbor, MI: University of Michigan Press.

Bodenhamer, D. J. (2010) "The potential of spatial humanities," in D. J. Bodenhamer, J. Corrigan and T. M. Harris (eds) *The Spatial Humanities: GIS and the Future of Humanities Scholarship*, Indiana, IN: Indiana University Press, pp. 14–29.

Bold, C., Knowles, R. and Leach, B. (2002) "Feminist memorializing and cultural counter-memory: the case of Marianne's Park," *Signs: Journal of Women in Culture and Society* 28(1): 125–7.

Brennan, J. (2012) *Taifa: Making Nation and Race in urban Tanzania*, Athens, OH: Ohio University Press.

Drucker, J., Lunenfeld, P., Presner, T. and Schnapp, J. (2012) *Digital Humanities*. Cambridge, MA: The MIT Press.

Chayda, J. (2003) "Mother politics: anti-colonial nationalism and the woman question in Africa," *Journal of Women's History* 15(3): 153–57.

Collins, P. H. (1998) *Fighting Words: Black Women and the Search for Justice*, Minneapolis, MN: University of Minnesota Press.

Coombes, A. E. (2003) *History after Apartheid: Visual Culture and Public Memory in a Democratic South Africa*, Durham, NC: Duke University Press.

Geiger, S. (1996) "Tanganyikan nationalism as 'women's work': life histories, collective bibliography and changing historiography," *The Journal of African History* 37(3): 465–78.

Geiger, S. (1997) *TANU Women: Gender and Culture in the Making of Tanganyikan Nationalism, 1955–1965*, Portsmouth, NH: Heinemann Press.

Horn, J. (2012) "Our Africa: mapping African women's critical resistance," Our Africa, October 16, 2012. Available online at: www.opendemocracy.net/5050/jessica-horn/our-africa-mapping-african-womens-critical-resistance (accessed June 4, 2013).

Hunt, N. R. (1989) "Placing African women's history and locating gender," *Social History* 14(3): 359–79.

Jaksch, M. (2013) "Feminist *Ujamaa*: transnational feminist pedagogies, community, and family in East Africa," in T. Jenkins (ed.) *Family, Community, and Higher Education*, New York: Routledge, 116–31.

Kerr, D. (1995) *African Popular Theatre: From Pre-colonial Times to the Present Day*, London: James Currey.

Kiisheweko, O. (2011) "Tanzania bids to document Africa's liberation struggle," December 11, 2011. Available online at: www.dailynews.co.tz/feature/?n=26337 (accessed June 4, 2013).

Kwan, M. (2002) "Feminist visualization: re-envisioning GIS as a method in feminist geographic research,"*Annals of the Association of American Geographers* 92(4): 645–61.

Lihamba, A. (2004) "Tanzania," in M. Banham (ed.) *A History of Theatre in Africa*, Cambridge, UK: Cambridge University Press, 233–46.

Lyons, T. (2004) *Guns and Guerrilla Girls: Women in the Zimbabwean Liberation Struggle*, Trenton: Africa World Press.

McClintock, A. (1993) "Family feuds: gender, nationalism, and the family," *Feminist Review* 44: 61–80.

McFadden, P. (2005) "Becoming postcolonial: African women changing the meaning of citizenship," *Meridians: feminism, race, transnationalism* 6(1): 1–18.

McLafferty, S. L. (2002) "Mapping women's worlds: knowledge, power and the bounds of GIS," *Gender, Place and Culture* 9(3): 263–9.

Mama, A. (2001) "Talking about feminism in Africa (interviewed by Elaine Salo in *Women's World*)," *Agenda* (African Feminisms I): 50.

Miller, K. (2009) "Moms with guns: women's political agency in anti-apartheid visual culture," *African Arts* 42(2): 68–75.

Miller, K. (2011) "Selective silencing and the shaping of memory: the case of the monument to the women of South Africa,"*South African Historical Journal* 63(2): 63–86.

Mlama, P. O. (1985) "Tanzania's cultural policy and its implications for the contribution of the arts to socialist development," *Utafiti* 7(1): 9–19.

Mohanty, C. (1991) "Under Western eyes: feminist scholarship and colonial discourses," in C. T. Mohanty, A. Russo and L. Torres (eds) *Third World Women and the Politics of Feminism*, Bloomington, IN: Indiana University Press, pp. 51–91.

Mollel, T. M. (1985) "African theatre and the colonial legacy: review of the East African scene," *Utafiti* 7(1): 20–9.

Morrow, S., Maaba, B. and Pulumani, L. (2004) *Education in Exile: SOMAFCO, the African National Congress school in Tanzania, 1978 to 1992*, Cape Town, SA: HSRC Press.

Mulligan-Hansel, K. (1995) "Tanzanian women's movement challenges structural and cultural limitations," *Feminist Voices* 8(2): 1.

Munslow, B. (1985) *Samora Machel: An African revolutionary*, London: Zed Books.

Ndziku, T. (1994) "Tanzanian women's roles in decision making policy," *Women in Action* 1 (94): 13.

Nieves, A. D. (2008) "Introduction: mapping geographies of resistance along the 16 June 1976 heritage trail," in A. K. Hlongwane (ed.) *Footprints of the 'Class of 76': Commemoration, Memory, Mapping and Heritage*, Johannesburg: The Library.

Nieves, A. D. (2009) "Places of pain as tools for social justice in the 'new' South Africa: black heritage preservation in the 'Rainbow' nation's townships," in W. Logan and K. Reeves (eds) *Places of Pain and Shame: Dealing with Difficult Heritage*, London: Routledge, pp. 198–214.

Nsekela, A. J. (1984) *Time to Act*, Dar es Salaam: Dar es Salaam University Press.

Nyerere, J. (1967) "The Arusha Declaration and TANU's policy on socialism and self reliance," Dar es Salaam: Tanzania, published by the Publicity Section, TANU, Dar es Salaam.

Perter, C. M. and Tenga, N. (1996) "The right to organize as the mother of all rights: the experience of women in Tanzania," *The Journal of Modern African Studies* 34(1): 143–62.

Pitcher, A. M. (2006) "Forgetting from above and memory from below: strategies of legitimation and struggle in postsocialist Mozambique," *Africa* 76(1): 88–112.

Plastow, J. (1996) *African Theatre and Politics: The Evolution of Theatre in Ethiopia, Tanzania and Zimbabwe – A Comparative Study*, Amsterdam: Rodopi.

Schroder (2012) *Africa after Apartheid: South Africa, Race and Nation in Tanzania*, Bloomington, IN: Indiana University Press.

Sheldon, K. (2010) *Historical Dictionary of Women in Sub-Saharan Africa*, Methuen, NJ: Scarecrow Press.

Shule, V. (2009) "Mwalimu Nyerere: The Artist," *Pambazuka News*, Issue 452, October 13, 2009. Available online at: www.pambazuka.org/en/category/features/59500 (accessed June 4, 2013).

Spivak, G. C. (1999) *Toward a History of the Vanishing Present*, Cambridge, MA: Harvard University Press.

Swai, E. (2010) *Beyond Women's Empowerment in Africa: Exploring Dislocation and Agency*, New York, NY: Palgrave Macmillan.

UNESCO (2005) "Roads to independence in Africa: the African liberation heritage project," UNESCO REPORT.

White, A. (2007) "All the men are fighting for freedom, all the women are mourning their men, but some of us carried guns: a race-gendered analysis of Fanon's psychological perspectives on war," *Signs* 32(4): 857–85.

Part III

Gender, the environment and community-based development

9 Gender, livelihoods and the construction of climate change among Masai pastoralists

Elizabeth Edna Wangui

Introduction

Scholarship on climate change adaptation recognizes that people modify their livelihoods based on a complex interaction of climatic and non-climatic stressors (Kirkbride and Grahn 2008; Nassef *et al.* 2009; O'Brien and Leichenko 2000; O'Brien *et al.* 2008). These approaches, however, provide little information about how communities themselves understand and construct climate change as part of this complexity of livelihood stressors. Yet the adaptation strategies that they choose have been linked to the climate knowledge that is available to them (Israel and Sachs 2013). Recent studies have added to our understanding of adaptive capacities by investigating how communities in Guanajuato, Mexico (Bee 2013) and the Old Peanut Basin, Senegal (Tschakert 2007) understand the causes and consequences of climate change.

This chapter contributes to the growing body of climate change literature by analyzing how pastoralists in three Masai communities in southern Kenya construct climate change through their gendered livelihoods. A more thorough understanding of pastoralist adaptation strategies and a deeper analysis of their capacity to adapt to climate change begins with an investigation of how "the adapters" construct climate change. I argue that social differences within pastoralist communities influence the construction of climate change by different individuals and groups and highlight the importance of gender in framing these constructions. Although it is important to examine different ways in which men and women construct climate change, it is crucial to recognize variations within each group and intersectionality with other factors. Gender analyses also indicate that men and women have shared experiences as a group, especially in the Masai communities under investigation, where gender responsibilities are still relatively well defined. This discussion adds to the literature on gender and climate change by demonstrating how gender responsibilities influence the construction of climate change at the local level. Broadly speaking, this diverse literature has drawn attention to men's and women's differential experiences with climate change (Dankelman 2010; Denton 2002; Lambrou and Piana 2006; Muthoni and Wangui 2013; Sultana 2010; Terry 2009). A thorough review of the literature on gender and climate change is not included in this chapter. However, it is important to note here that pastoral livelihoods remain under-represented in the literature examining the role that gender plays in climate adaptation. This chapter joins Omolo's (2010) research in northern Kenya as one of the few existing studies that address this gap.

Constructing climate change

In recent decades, investigations into the social construction of the natural world have been of significant interest to social scientists (Baker and Bridge 2006; Castree and Braun 1998; Merchant 1980). This diverse literature offers the idea of construction as philosophical critique, suggesting the socially situated position of knowledge and knowledge production (Demeritt 2002). This view overlaps with feminist critiques of science that question objectivity, truth and reality and argue for the privilege of the partial perspective whereby an individual's knowledge is mediated by their position and identity (Braidotti *et al.* 1994; Haraway 1991; Harding 2004; Seager 2009). Demeritt (2001) specifically examines the production of the knowledge of global warming and concludes that climate scientists' understanding of this process is inseparable from the social practices within which climate science happens. Furthermore, Israel and Sachs (2013) draw on existing feminist critiques of science to advocate for a rethinking of climate change. This rethinking involves a "reconstruction of climate science [that sees] climate models and their projection maps as one form (among many) of technologically mediated, embodied, and situated knowledge about the climate" (Israel and Sachs 2013: 43).

Climate scientists provide an important, albeit partial, understanding of climate change. As illustrated by Forsyth and Walker (2008) and Beymer-Farris and Bassett (2012), concerns guided by such a partial understanding of climate change caused some major conservation agencies to preserve ecosystem health with little regard for local communities, whose livelihoods are impacted by such efforts. The two studies illustrate how a partial understanding of environmental change can lead to the development of environmental narratives that construct local people as destroyers of the environment and their livelihoods as illegal. A more comprehensive understanding comes from the inclusion of alternative positions to such discussions, including views from the local scale. Ultimately, it is at the local level where impacts of climate change are felt and adaptations practiced. If we are to arrive at responsible climate policies that are sustainable and equitable, I argue that information coming from climate scientists needs to be taken together with perspectives and experiences from local communities. This is in line with recent calls from feminist scholars to support efforts that "re-value the local environmental knowledges that have been thoroughly marginalized in climate change discourse" (Israel and Sachs 2013: 35). Israel and Sachs add that such rethinking of climate science can "create space for new knowledges about climate, knowledges that come out of different social (and geographic) locations and from different tools than mainstream climate science" (2013: 46).

The local scale has also been emphasized in the context of climate justice by advocating "putting the most vulnerable first" (Paavola and Adger 2006: 604). This is not to say that the local scale provides information that is necessarily superior to other scales. There is, however, much to be gained from shared conversations across different scales. This chapter adds to critical work on climate change and communities that emphasizes the role of local and community scales by laying the groundwork for these conversations (Leichenko and O'Brien 2002; Tschakert 2007). For example, Tschakert's research on how rural communities in Senegal understand climate variability and change includes a discussion of knowledge from climate science. By focusing on the local scale, at which pastoralists produce their livelihoods in three Masai communities in Kenya, I argue that the concept of climate change is constructed differently by men and women. The position that climate change is a socially constructed and contingent process does not deny the materiality of climate change. The discussion in this chapter emphasizes that people's lived experiences have a direct bearing on how they construct climate change, and these constructions will differ between men and women based on their social differences and societal roles.

Researching gender and climate change

The research in this chapter was part of a larger multidisciplinary project that examined the dynamics between coupled human-biophysical systems in savannas under climate change. The research included an investigation of how climate-driven changes impacted livestock and crop production as well as peoples' livelihoods in Kenya and Tanzania. The results reported in this chapter are based on national climate data collected from published sources and fieldwork conducted in three Masai communities, in Kajiado South Constituency in southern Kenya (Figure 9.1). The three communities, Empiron, Mbirikani and Risa, were selected based on their main livelihood activities as identified by team members who had field experience in these communities dating back to the late 1970s. Collectively, these communities represented the main livelihood trajectories for pastoralist communities in southern Kenya. In Empiron, pastoralism included both irrigated and rain-fed farming. In Mbirikani, pastoralism included irrigated farming and some additional incomes obtained through wildlife conservation. Risa had the highest reliance on livestock among the three communities, but people also practiced irrigated farming. None of the Masai communities in Kajiado South Constituency relied solely on livestock for their livelihood.

Three focus group discussions were held in each of the three communities. Participants were selected in collaboration with research assistants and other opinion leaders from the communities. Care was taken to achieve a balance between knowledge on the topics under discussion and diversity in age, education and wealth status. Sixteen participants were invited to each meeting and attendance ranged from eight to fourteen. The first meeting was an event history calendar discussion (Caspi *et al.* 1996). An equal number of men and women were invited to each event history meeting. Event history calendars were developed by researchers

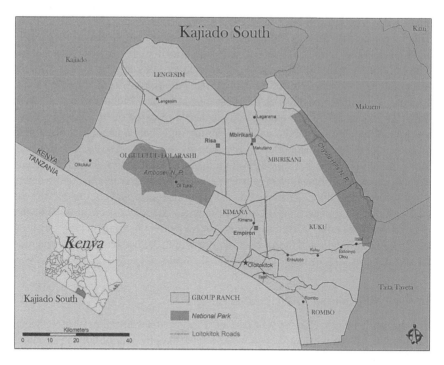

Figure 9.1 Case study sites in Kajiado South Constituency, Kenya. (Fieldwork by Elizabeth Edna Wangui, 2008.)

interested in life histories (e.g. Freedman *et al.* 1988) and rely on temporal and thematic cues that work together to produce high quality retrospective reports even after long retention intervals (Belli 1998). At each meeting, participants were asked to talk about major climatic and non-climatic occurrences of significance to their livelihoods. Participants were encouraged to go as far back as they could remember to establish a starting point. Events (climatic and non-climatic) mentioned were often cultural time markers that we were able to translate to a Western calendar. This translation paved the way for discussions of not only change, but also when change happened, in subsequent meetings. For example, participants would include cultural events such as "circumcision of the *ilkimunyak* age set." In that particular community, this event happened in 1980. There were enough cultural time markers mentioned, that it was not difficult to translate climatic events cited to a Western calendar. For each climatic event mentioned (e.g. "terrible drought"), participants were asked to discuss the specific impact on their livelihood, what different people did to mediate the impact and for how long they did it. Participants also drew comparisons between different climatic events as a means of evaluating how impacts on their livelihoods may be changing.

Event histories were followed by single gender meetings where participants made maps of their livelihood space (Edmunds *et al.* 1995) and discussed the changes that these spaces had undergone. Participants at each meeting were asked to talk about the livelihood activities they perform and indicate where they go to perform these activities. In the course of the discussions, participants decided what to include on the map, where on the map to place spatial markers and the spatial coverage the map should have. They were asked to talk about their experiences with the landscape, the changes they had observed at the places they indicated on the map, when they first noticed the changes, the progression the changes took and to what they attributed the changes. The mapping exercise revealed the gendering of both pastoralist livelihood space and people's quotidian experiences of the physical environment. It further revealed how spaces are valued and how men and women relate differently to particular resources. Men and women frequently possess different environmental knowledges, based on their different daily experiences (Braidotti *et al.* 1994; Rocheleau *et al.* 1996), which is what the study sought to capture. The interviews were later translated, transcribed and analyzed using Nvivo qualitative data analysis software. All names used in this chapter are pseudonyms.

Climate variability and local narratives in Kenya

In this section, I provide an overview of Kenya's climate based on data available at national scales and examine potential complementarities between national climate data and local climate narratives. Kenya's rainfall is primarily influenced by the movement of the inter-tropical convergence zone, creating two distinct wet periods. The long rains fall between March and May and the short rains between October and December. In total, the two rainy seasons produce a monthly average of 50–200 mm. The Stockholm Environmental Institute (SEI) emphasizes that there is considerable spatial variation in onset, duration, intensity and amount of rainfall (with some places receiving as much as 300 mm per month (SEI 2009)). A UNDP report by McSweeny *et al.* (2010) provides an analysis of Kenya's climate based on available gridded station data (1960 to 2006) and future climate projections. The report indicates that mean annual temperatures have increased by 1 degree Celsius since 1960 and are projected to increase by 1–2.8 degrees by 2050 and by 1.3–4.5 degrees by 2090. Rainfall shows no statistically significant trends at 95 percent confidence limits. However, monthly rainfall totals show a decline between 1960 and 2006, with the largest decrease being observed during the long rains (−3.9 percent per decade). An increasing proportion of rain falls during heavy events, though this is not statistically significant. Model projections do, however, consistently indicate an increase in these heavy rain events, ranging from 1 to 13 percent by 2090. Overall, rainfall in Kenya is predicted to change by −1 to 48 percent by 2090. Norrington-Davies and Thornton (2011) state that drought frequency will likely remain the same as it is now, but the rise in temperature will increase the intensity of the droughts. However, they add that a large disparity in the model projections of the amplitude of future El Niño events exists. Since El Niño events influence drought events in the region, the disparity adds uncertainty to current drought projections.

Country level climate analysis is useful, as it provides continuous quantitative information on past and projected trends that could inform ground level adaptation priorities. Considerable uncertainty exists, however, in the wide range of projected increases in rainfall and temperature. Additionally, even with downscaled climate analysis, spatial resolution remains coarse. For example, the UNDP report (McSweeny *et al.* 2010) uses a 0.5 degree by 0.5 degree grid (55.5 km by 55.5 km) for the observational data and a 2.5 degree by 2.5 degree grid for the projected data. This means that in areas where elevation (and hence rainfall and temperature) changes rapidly, as it does on the lower slopes of Mt. Kilimanjaro, national and regional data fail to capture much of this variation.

Additionally, climatic variability is not fully addressed in country level analysis. Yet it is an important way that long-term climate change manifests itself and thus forms a critical challenge to livelihoods (Smucker and Wisner 2008). Variability is especially high in arid and semi-arid areas, which is where most pastoralism is practiced, and is indicated by local observations of more frequent and more intense and erratic weather patterns. Weather patterns are characterized by cycles of drought and devastating annual flooding, which have become more severe since 2000 (Norrington-Davis and Thornton 2011). The cost of such erratic weather is high. The SEI estimates that Kenya experiences a flood that costs about 5.5 percent of GDP every seven years, and a drought that costs about 8 percent of GDP every five years (SEI 2009). While the SEI acknowledges that future costs of climate change are uncertain, they argue that the economic costs of climate change in Kenya could be equivalent to a loss of 2.6 percent of GDP each year by 2030.

The information presented above highlights the need to supplement national climate data with local narratives to obtain a more comprehensive picture of Kenya's climate. Local climate narratives that emerge from event histories, and from the creation of livelihood maps, add to data that is available from smaller scales to better inform climate change adaptation concerns. First, the narratives were collected from communities that occupy two different agro-ecological zones within 35 km of each other (Figure 9.1). The narratives could therefore add spatial detail to the national data, which averages climate information over a large area as explained above. For example, Empiron, which lies in the wetter agro-ecological zone, did not report a drought in 1974. In contrast, people in Mbirikani and Risa communities in the drier agro-ecological zones talked about the failure of the rains in 1974. Similarly, Empiron is a highland community that did not report the 1997/8 El-Niño floods. The other two communities occupy low-lying areas and both reported flooding during the El-Niño. Second, people spoke about temporal variability that would be difficult to capture in the national data. National rainfall and temperature data are available on a monthly basis. Variations that occur within a month are important for the crop and livestock calendar and form the basis on which people make decisions that directly impact their livelihoods. Farmers who rely on rainfall spoke about increasing variability in the onset of the rains, which makes it challenging to plant at the optimal time. Narratives of temporal variability could add detail to available national data. Third, climate narratives reveal a clear co-dependence of different livelihood activities, which can mediate the impact of a climatic event. For example, the 2004 drought led to massive crop failure in all three communities; however, livestock benefitted from these as the failed crops were an important source of fodder.

Fourth, because of the integrated nature of climate and livelihoods, discussions of climatic events occur simultaneously with discussions of long-term adaptation and immediate coping strategies. Through such accounts, the evolution of coping to adaptation, the negotiations involved in that evolution and the gendered labor demands of adaptation become clear. Finally, livelihood changes occur in response to multiple stressors. People's constructions of climate change are therefore complex and not based on simple rainfall and temperature accounts. Understanding their adaptive capacities depends heavily upon this construction of climate change.

Livelihoods among Masai pastoralists

Pastoralists in sub-Saharan Africa occupy dryland areas, which have historically been politically distant from central governments (Anderson and Broch-Due 1999; Kirkbride and Grahn 2008). Combined with the effects of climate change, the poverty levels pastoralists face are expected to compound their development challenges in the coming years (Kirkbride and Grahn 2008; Nassef *et al.* 2009; Wangui *et al.* 2012). This study focuses on three Masai communities, Empirion, Mbirikani and Risa, located in Kajiado South Constituency in southern Kenya (Figure 9.1). Empiron is part of the recently subdivided Kimana Group Ranch, which is located in the midland zone and receives enough rainfall to support rain-fed farming; however, it is too dry to support a crop harvest each year. Frequent crop failures means that irrigation supplements rain-fed farming and that most people who live here do not rely solely on rainfall. A few pastoralists in this community raise exotic cows that provide a higher milk yield. The high labor and financial costs, as well as the susceptibility to livestock disease, limit the keeping of exotic cattle.

The other two communities are located in the driest agro-ecological zone in the constituency. The Mbirikani community is part of the Mbirikani Group Ranch, which partnered with wildlife conservation organizations to provide employment for community members and scholarships for local primary and secondary schools. Almost all members of this community have small plots of farmland in the irrigated parts of the ranch (Figure 9.2). The Mbirikani Masai are increasingly keeping non-Masai livestock breeds, which produce more meat than local Masai cattle but are also less drought tolerant. Finally, Risa, the smallest of the three communities, relies heavily on pastoralism. This community is part of the Olngulului-Lolarashi Group Ranch, which has partnerships with wildlife conservation organizations working in the Amboseli area. Unlike Mbirikani, most of the benefits from the partnership do not reach the people of Risa. Almost all of the homesteads in this area have small plots of farmland in the irrigated zone. Additionally, in a regular dry season, Risa families migrate with their livestock to dry season homes located within the group ranch. School-going children

Figure 9.2 Irrigated farming in Mbirikani Group Ranch. (Photo by Elizabeth Edna
Wangui, 2008.)

remain behind with a few adults who take care of them. A few livestock also
remain in the village to provide milk and other necessities for those who do not
migrate.

In all three Masai communities, livelihoods are changing as pastoralism diver-
sifies to include farming and other activities (Campbell *et al.* 2000). As a result,
clearly defined, but constantly shifting gender roles are evident (Wangui 2008,
2012). Overall, livestock activities that have been dominated by men in the past are
now predominantly done by women. However, day-to-day decisions are still made
by men, since they own the livestock. For the most part, men's labor has shifted
from grazing to cash crop farming. Like other parts of Masailand, Kajiado South
is experiencing land subdivision from communal to individual tenure (Mwangi
2007). This subdivision is in different stages in the three communities studied. In
Empiron, land subdivision has been completed and people have begun to sell their
(now) private land or fence it for new uses, such as wildlife conservation (Kiiru
2012). In Mbirikani, land subdivision was expected, but had not begun at the time
of this fieldwork.

Gendered constructions of climate change

The data collected in this research reveals that people's constructions of climate
change are built from direct experience and interpreted through their local knowledge systems. Three interrelated themes emerge from this research. First, when
they talk about their livelihood spaces and how they experience change in these
spaces, the people reveal *how they know* about climate change and the ways in
which these ways of knowing are gendered. Second, they provide information on
what they know about climate change and the degree to which this is important to

their livelihoods. Third, they demonstrate *how they act* on the knowledge to adapt their livelihoods to the changes they experience.

Men and women's knowledge of climate change is deeply tied to their experiences with the physical and social environments, through their gendered responsibilities. For women, this experience stems from gathering and grazing, while for men it is primarily based on grazing. These activities take them to different points on the landscape and, over time, they observe changes in the physical environment that have a direct bearing on their gendered responsibilities. Discussions clearly reveal that constructions of climate change are influenced by a comprehensive analysis of what climate change means to men and to women. These discussions also take into consideration how a changing climate impacts the physical environment as well as their gendered responsibilities. As demonstrated in this analysis, at the local level, climate change is not constructed simply as a process in the physical environment that can be described in terms of rainfall and temperature.

Local pastoralists often view climate change as a new challenge to their grazing and gathering journeys. Since the livelihood practices of Masai pastoralists are heavily dependent on vegetation, one of their first experiences with climate change is through its effect on vegetation. Therefore, their constructions of climate change often begin with discussions about the availability of plants on which they depend for household activities, such as construction, gathering firewood, grazing livestock and storing food items. Both men and women in Kajiado South talked about how climate change and increasing climate variability negatively impact the plants that they need to for their livelihoods (Table 9.1). Distinct gendered patterns are evident in how men and women talk about plants. Men in all three communities discuss the changes that have occurred to plant species that are important in grazing. Women, on the other hand, view the same spaces not just as places to graze but also as places to collect firewood, material for house construction and plants used in the calabashes used for milk storage. Women therefore tend to have a much broader understanding of changes in plant species than men (Table 9.1).

Women's journeys in search of plant materials have changed, and the new places tend to be further away from their homes. These new routes create new forms of human-wildlife conflict, different from those previously reported in literature (e.g. Campbell *et al.* 2000). Older women explained that they have to rely more heavily on their daughters and daughters-in-law to collect firewood for them in light of these new conflicts with wildlife. This reliance on the labor of younger women and girls adds a new dimension to the relationships and interactions between women of different generations. The women explained that in some families this has the potential to change power relations between mothers-in-law and daughters-in-law, which, as reported in other parts of the world (Harris 2006), is usually skewed in favor of the mothers-in-law.

Women construct climate change in terms of new demands on their labor that also influence relations with their husbands. For example, women talked about how the changing climate combined with increasing monetary demands to negatively impact their labor and particularly the time they use to rest and

Table 9.1 Gender and livelihood uses of plants negatively impacted by climate change in Kajiado South, Kenya

Masai name	Botanical name	Use*	Mentioned by	Location**
Emangulai	Grewia villosa	H	Women	EMR
Endimonyua	Chloris virgata	G	Men, women	M
Enkamaloki	Cadaba falinosa	H	Women	R
Enkamologe	Cadaba falinosa	H	Women	EMR
Entimonyua	Cenchurus ciliaris	Gc	Men	EM
Erikaru	Digitaria macrocephala	Gc	Men, women	EMR
Esiteti	Grewia bicolor	H	Women	EMR
Oiti	Acacia mellifera	CFH	Women	EMR
Olamoloki	Cadaba varenosa	F	Women	EM
Olchurrai	Acacia ancistroclada	FH	Women	EMR
Oldebe	Acacia nubica	F	Women	R
Olmapite	Cordia sinensis	CFH	Women	EMR
Oloileroi	Boscia angustifolia	FH	Women	EMR
Olorien	Rhynchosia minima	C	Women	E
Oltepesi	Acacia tortilis	F	Women	EMR
Oltontolian	Blepharis integrifolia	G	Men	R
Olungoswa	Balanites aegypticum	FH	Women	EMR
Orbutiani	Plectranthus sp	Gs	Men	E
Oseki	Cordia goetzei	CH	Women	E

Source: fieldwork by Elizabeth Edna Wangui, 2008.

Notes: often the same Masai plant name can be used to refer to several different plants. The Masai plant names provided by participants during data collection are used here. A translation of the Masai plant names to botanical names was provided by a botanist who participated in this fieldwork.

*Plant uses are coded as follows: C is calabash used to store milk, F is firewood, G is grazing, Gc is grazing cattle, Gs is grazing sheep and H is house building.

**Location is coded as follows: E—Empiron, M—Mbirikani and R—Risa.

socialize. This quote by a woman from Mbirikani captures this sentiment of overcommitted labor:

> Women used to milk, do beadwork, sleep a lot always around the home. Now you can't find a woman sleeping during the day. We no longer work around our homes because these days everything depends on money and you must leave your house to get money. . . as the bible says, if you depend on someone else, you will be cursed. So you must depend on yourself. But we are happy we have more work, because these are opportunities.
>
> (Naserian, women's meeting, Mbirikani)

Her statement also reflects a dimension of changing gender relations that was evident in all our meetings with women. Naserian's comment about the need to "depend on yourself" rather than someone else indicates changing gender responsibilities and relations exemplified by increases in women's labor and monetary contributions within the household. A detailed examination of this process is documented elsewhere (Wangui 2008, 2012). As illustrated in the quotes below, changing gender relations are not without some tensions in the communities of Empiron and Mbirikani:

> Men don't like us being clean and going out and becoming independent. Men think women have become proud and are ignoring their husbands and respecting them less. Men feel challenged to measure up to their wives' ability to buy food and they also go out and buy food. In the eyes of the men, a good wife is one who does not change and develop.
>
> (Nabulu, women's meeting, Empiron)

> Climate change has affected our livelihood more than other factors and our level of activity has had to increase. But the level of activity for most men has not increased, just the traders. Traders have no time to rest. The rest of the men sleep and drink. Maybe now it is our turn to dominate. Things are turning upside down. The head used to be the husband, the wife was the neck. Now it is upside down. But the men still own property.
>
> (Nasieku, women's meeting, Mbirikani)

A "clean" woman is one who is about to embark on an activity in a place that she does not usually occupy. The term is symbolic of a woman's readiness to take on responsibilities that she did not previously have. Women explained that they are assuming financial responsibilities previously associated with men. Nasieku's comment about 'things turning upside down' reinforces what sometimes happens when opportunities are provided by women's groups. When they first started forming groups in Empiron and Mbirikani, women faced challenges as their husbands would restrict their movements and hence their ability to attend group meetings (Wangui 2004). These restrictions were eventually lifted when men started to see household benefits of these women's groups.

Men and women's constructions of climate change are interwoven with discussions of socio-political processes that challenge adaptation strategies. In both meetings at Empiron, men and women talked at length about how recent group ranch subdivision will impact pastoralism as they practice it. Subdivision, a change in land tenure from group ownership to individual titles, is an ongoing process in Masailand and has significant implications for livestock mobility (Mwangi 2007). Men are concerned that many pastoralists are leasing or selling their (now) private parcels of land to tourism and related wildlife conservation efforts. This is seen as a loss of grazing areas and an increase in wildlife-related conflicts, which further challenges adaptation.

A loss of communal grazing is part of the reason that pastoralists of Empiron have already reduced their herds and diversified their livelihoods. One man at the meeting expressed concern that "in the end, there will be no livestock. Where will you graze it? If you have private land for two cows, you will have to keep just two cows." Women also talked about how group ranch subdivision creates new challenges for their responsibilities. Many women see land subdivision as a process that combines with climate change to put further pressure on access to firewood and pasture. Poor women in particular say that now they either have to beg for plant material to build their houses or travel very far. They feel that only rich people will gain from subdivision. One woman at the meeting summarized this market logic of resource access in her comment that "they [the rich members of the community] will be lucky since they will pay more money to get more resources and this will make it even harder for the women." The women said that they had stopped fencing their homesteads since they could no longer find enough plant material to surround the entire homestead. What little plant material they find is used to fence livestock kraals. Unfenced homesteads are more likely to experience human-wildlife conflict.

Mitigating the impact of climate change

Masai pastoralists have put in place several livelihood strategies to mitigate the impact of climate change in grazing and gathering journeys. The first strategy to mitigate climate change that men spoke about was a land use management scheme that covers four group ranches in the constituency. Risa and Mbirikani lie in two of these group ranches and are involved in the grazing management scheme. The scheme was put in place by local group ranch officials after the devastating 1984 drought in order to conserve part of the rangeland for grazing during the dry season. An area locally referred to as *olopololi* has also been set aside, where firewood can be collected and where livestock that does not migrate during the dry season can graze. There are also places set aside for settlement. Risa and Mbirikani have had different levels of success with the scheme. In Risa, people's assessment of the process was positive as the policies of use set by the group ranch officials are respected. In Mbirikani, however, where group ranch subdivision is expected and national and group ranch politics intertwine, people are abandoning the scheme and using the conserved area before it is officially open for grazing. Since 2000, many group ranch politicians align themselves with national political parties. When one set of group ranch officials wins the local election, some local group ranch members from the opposition challenge the group ranch rules of land use by building houses in the conserved area and *olopololi*. These measures are in anticipation of being allocated these preferable locations once subdivision is underway. This open contestation is a clear illustration of the complexity of the socio-political context within which climate change adaptation happens. Complex interactions of national and local politics overshadow attempts at planning for climate change, leading to a breakdown of local institutions that would mitigate the impact of climate change. Climate change has made some parts of the group ranch even more desirable and any group that fears being left out during the subdivisions (since they are not in power) is inclined to move into these desirable places.

The second strategy to mitigate the impact of climate change on livelihoods is that women either walk further in search of the desired firewood species (Risa) or they turn to alternative, less desirable trees in the face of diminishing firewood. Less desirable trees produce firewood that burns too fast or produces too much smoke. They also harvest trees that were previously protected by cultural taboos. Finally, some women find new ways of obtaining plants used in calabashes that were formerly found close to their homes. The women from Empiron live within walking distance of Loitokitok, the main trading center in the constituency. Those who can afford it obtain *olorien* products from the Loitokitok market, whereas those who cannot afford it use *oseki* as an alternative (Table 9.1). Others rely on their social networks to obtain plants harvested from the distant Chyulu Hills (Figure 9.1).

Discussions of problems with water did not feature as prominently as was expected during the meetings. When people were asked about this, they explained that the water situation has improved due to recent government interventions. In Mbirikani, they attributed this to the installation of the Nolturesh pipeline, which takes municipal water from springs near Loitokitok town to towns outside the constituency (Figure 9.1). The pipeline passes through the group ranch and people are able to obtain water from taps installed along it (Figure 9.3). People also have more water storage tanks than in the past for rain water harvesting. In Risa, women said that they have several water collection and livestock watering points, one of which has been built in the last five years using money allocated by the central government. Men in Risa also said that the water situation has improved, especially in places where calves, sheep and goats go for water during the dry season. Challenges remain, however, as members of the community sometimes need to contribute money to buy gas for water pumps in some water access points.

In sum, the last two sections demonstrate how knowledge and experience of climate change is mediated by gender through social roles and responsibilities. This mediation produces different constructions of climate change that are linked to gendered livelihoods. The different positions occupied by men and women each

Figure 9.3 Water tap located along the Nolturesh pipeline. (Photo by Elizabeth Edna Wangui, 2008.)

provide a privileged partial perspective (Haraway 1991) which become complementary when taken together. This discussion also supports Demeritt's (1998;
2001) argument that climate change is not just a statistical abstraction, and the
social context within which climate scientists operate influences their understanding of climate change. Likewise, the socio-political contexts of Kajiado South
influence how climate change is experienced and therefore constructed by Masai
pastoralists. Thus, land tenure, gendered responsibilities and relations and the
intersection of national and local politics impact the construction of climate
change.

Summary and concluding remarks

Although researchers have argued for downscaled climate models to directly
inform adaptation (Ziervogel and Zermoglio 2009), this research shows the need
to also incorporate local understanding of climate change in efforts to help communities adjust to negative impacts on their livelihoods. Such conversations would
occur across different scales and have the potential to inform climate policies that
are both sustainable and equitable. Discussions from the local scale also reveal
climate change as a social process that requires new tools and insights to fully
comprehend. People making adaptation decisions have to take a lot more into
consideration than rainfall and temperature when they make those decisions.

People's constructions of climate change are primarily linked to knowledge
acquired in the process of practicing their gendered responsibilities. In addition to
rainfall and temperature, constructions of climate change incorporate other aspects
relevant to grazing and gathering journeys. Gendered labor demands (such as for
grazing and gathering fuelwood) and changing gender relations form critical parts
of people's constructions of climate change that demonstrate the need to consider
a complex biophysical and social process along with other stressors when making
adaptation decisions. This idea mirrors Hulme's (2007) contention that climate
change is not only a projection of future physical climate change, but also of
people's perception, understanding and responses to local climate variability.

Examining climate change through a constructionist and gender lens is useful in
understanding what it means to pastoralists (Denton 2002; Omolo 2010; Aguilar
2010). Field discussions of "climate change" are challenged by the academic
abstractions inherent in the concept. When they talk about climatic elements, people tend to focus on short-term climatic variability, as this is the aspect of climate
change that they have experienced in their lifetimes. This tendency is evident in
the event history calendars constructed by the three Masai pastoralist communities
participating in this study. Entering this discussion through a constructivist lens
and starting from the communities' lived experiences is useful as it allows us to
overcome this challenge of the temporal scale whereby people experience variability over the short term, yet climate changes over a long time. This approach also
has the potential to allow for a more thorough understanding of how people make
decisions about adaptation, since such activities are based on a complex analysis
of both the social and biophysical context. As is illustrated by climate research

that focuses more on communities, people respond to multiple stressors and not just climate change (O'Brien and Leichenko 2000).

The constructionist lens further reveals climate change as socially produced and contingent on local contexts. Including both men's and women's voices in this study captures multiple interpretations of climate change as each group focuses on the spaces and changes they experienced first-hand. This information, taken together with constructions from climate scientists, provides a more complete understanding of climate change, current adaptation and possible adaptation futures. The partial perspectives (Haraway 1991) revealed in this chapter are important alternatives to ongoing discussions within major conservation agencies, where local communities are seen as a hindrance to climate change adaptation and mitigation (Beymer-Farris and Bassett 2012; Forsyth and Walker 2008).

Finally, this research has implications that go beyond responding to climate change to address the broader issue of promoting responsible development programs. Responding to climate change and other development challenges requires utilizing knowledge systems from multiple scales and sectors and crossing epistemic divides. Given the power imbalance that prioritizes scientific over local knowledge systems, navigating such epistemic divides would be challenging, but needs to be embraced.

Acknowledgements

This research has been funded as part of the National Science Foundation Biocomplexity of Coupled Human and Natural Systems Program, Award No. BCS-0709671. I would like to thank the men and women of Empiron, Mbirikani and Risa for taking the time to share their knowledge with me. Simon Mugatha helped translate Masai plant names to their botanical names. I am grateful to Ann Oberhauser, Tom Smucker and two additional reviewers for their insightful and constructive feedback on earlier drafts of this chapter.

References

Aguilar, L. (2010) "Establishing the linkages between gender and climate change adaptation and mitigation," in I. Dankelman (ed.) *Gender and Climate Change: An Introduction*, London: Earthscan, pp. 173–93.

Anderson, D. and Broch-Due, V. (1999) *The Poor Are Not Us: Poverty and Pastoralism in Eastern Africa*, Athens: Ohio University Press.

Baker, K. and Bridge, G. (2006) "Material worlds? Resource geographies and the 'matter of nature'," *Progress in Human Geography* 30(1): 5–27.

Bee, B. (2013) "Who reaps what is sown?: A feminist inquiry into climate change adaptation in two Mexican *ejidos*," *ACME: An International E-Journal for Critical Geographers* 12(1): 131–54.

Belli, R. F. (1998) "The structure of autobiographical memory and the event history calendar: potential improvements in the quality of retrospective reports in surveys," *Memory* 6(4): 383–406.

Beymer-Farris, B. and Bassett, T. J. (2012) "The REDD menace: resurgent protectionism in Tanzania's mangrove forests," *Global Environmental Change* 22: 332–41.

178   *E. E. Wangui*

Braidotti, R., Charkiewicz, E., Hausler, S. and Wieringa, S. (1994) *Women, the Environment and Sustainable Development: Towards a Theoretical Synthesis*, London: Zed Books in association with INSTRAW.

Braidotti, R., Charkiewicz, E., Hausler, S. and Wieringa, S. (1999) "Response to drought among farmers and herders in southern Kajiado District, Kenya: a comparison of 1972–1976 and 1994–1995," *Human Ecology* 27(3): 377–416.

Campbell, D. J., Gichohi, H., Mwangi, A. and Chege, L. (2000) "Land use conflict in Kajiado District, Kenya," *Land Use Policy* 17(4): 337–48.

Caspi, A., Moffitt, T. E., Thornton, A., Freedman, D., Amell, J. W., Harrington, H., Smeijers, J. and Silva, P. A. (1996) "The life history calendar: a research and clinical assessment method for collecting retrospective event-history data," *International Journal of Methods in Psychiatric Research* 6: 101–14.

Castree, N. and Braun, B. (1998) "The construction of nature and the nature of construction: analytical and political tools for building survivable futures," in B. Braun and N. Castree (eds) *Remaking Reality: Nature at the Millennium*, New York: Routledge, pp. 2–41.

Dankelman, I. (ed.) (2010) *Gender and Climate Change: An Introduction*, London: Earthscan.

Demeritt, D. (1998) "Science, social constructivism, and nature," in B. Braun and N. Castree (eds) *Remaking Reality: Nature at the Millennium*, New York: Routledge, pp. 172–92.

Demeritt, D. (2001) "The construction of global warming and the politics of science," *Annals of the Association of American Geographers* 91(2): 307–37.

Demeritt, D. (2002) "What is the 'social construction of nature'?: a typology and a sympathetic critique,' *Progress in Human Geography* 26(6): 767–90.

Denton, F. (2002) "Climate change vulnerability, impacts, and adaptation: why does gender matter?," *Gender and Development* 10(2): 10–20.

Edmunds, D., Thomas-Slayter, B. and Rocheleau, D. (1995) "Gendered resource mapping: focusing on women's spaces in the landscape," *Cultural Survival Quarterly* 18(4): 62–8.

Forsyth, T., and Walker, A. (2008) *Forest Guardians, Forest Destroyers: The Politics of Environmental Knowledge in Northern Thailand*, Seattle, WA: University of Washington Press.

Freedman, D., Thornton, A ., Camburn, D., Alwin, D. and Young-DeMarco, L. (1988) "The life history calendar: a technique for collecting retrospective data," *Sociological Methodology* 18: 37–68.

Haraway, D. J. (1991) *Simians, Cyborgs and Women: The Reinvention of Nature*, London: Free Association Books.

Harding, S. (ed.) (2004) *The Feminist Standpoint Theory Reader: Intellectual and Political Controversies*, New York: Routledge.

Harris, L. M. (2006) "Irrigation, gender, and social geographies of the changing waterscapes of southern Anatolia," *Environment and Planning D: Society and Space* 24: 187–213.

Hulme, M. (2007) "Geographical work at the boundaries of climate change," *Transactions of the Institute of British Geographers* 33(1): 5–11.

Israel, A. L., and Sachs, C. (2013) "A climate for feminist intervention: feminist science studies and climate change," in M. Alston and K. Whittenbury (eds) *Research, Action and Policy: Addressing the Gendered Impacts of Climate Change*, Dordrecht: Springer, pp. 33–51.

Kiiru, W. W. (2012) "Understanding the spatial, temporal and socio-economic factors affecting human-elephant conflict around Amboseli National Park in Kenya," Ph.D. Dissertation, University of Kent, Canterbury, Kent, UK.

Kirkbride, M. and Grahn, R. (2008) "Survival of the fittest: pastoralism and climate change in East Africa," Oxfam Briefing Paper No. 116. Available online at: www.oxfam.org/policy/bp116-pastoralism-climate-change-0808 (accessed May 19, 2013).

Lambrou, Y. and Piana, G. (2006) "Gender: the missing component in the response to climate change," Rome: Food and Agricultural Organization. Available online at: www.fao.org/sd/dim_pe1/docs/pe1_051001d1_en.pdf (accessed May 19, 2013).

Leichenko, R. and O'Brien, K. (2002) "The dynamics of rural vulnerability to global change: the case of southern Africa," *Mitigation and Adaptation Strategies for Global Change* 7: 1–18.

McSweeney, C., New, M., Lizcano, G. and Lu, X. (2010) "The UNDP climate change country profiles: improving the accessibility of observed and projected climate information for studies of climate change in developing countries," American Meteorological Society. Available online at: http://journals.ametsoc.org/doi/pdf/10.1175/2009BAMS2826.1 (accessed May 19, 2013).

Merchant, C. (1980) *The Death of Nature: Women, Ecology and the Scientific Revolution*, San Francisco: Harper.

Muthoni, J. and Wangui, E. E. (2013) "Women and adaptive capacity to climate change: opportunities and constraints in Mangio Village, Tanzania," *African Geographical Review* 32(1): 59–71.

Mwangi, E. (2007) "Subdividing the commons: distributional conflict in the transition from communal to individual property rights in Kenya's Masailand," *World Development* 35(5): 815–34.

Nassef, M., Anderson, S. and Hesse, C. (2009) "Pastoralism and climate change: enabling adaptive capacity," London: Overseas Development Institute. Available online at: www.odi.org.uk/resources/details.asp?id=3304&title=pastoralism-climate-change-adaptation-horn-africa (accessed May 19, 2013).

Norrington-Davies, G. and Thornton, N. (2011) "Climate change financing and aid effectiveness: Kenya case study," OECD. Available online at: www.oecd.org/countries/kenya/48458443.pdf (accessed May 19, 2013).

O'Brien, G., O'Keefe, P., Meena, H., Rose, J. and Wilson, L. (2008) "Climate adaptation from a poverty perspective," *Climate Policy* 8: 194–201.

O'Brien, K. L., and Leichenko, R. (2000) "Double exposure: assessing the impacts of climate change within the context of economic globalization," *Global Environmental Change* 10: 221–32.

Omolo, N. A. (2010) "Gender, pastoralism and climate change: vulnerability and adaptation in northern Kenya," Final Technical Report to International Development Research Center and Department for International Development. Available online at: http://start.org/download/accfp/omolo-final.pdf (accessed May 19, 2013).

Paavola, J. and Adger, W. N. (2006) "Fair adaptation to climate change," *Ecological Economics* 56: 594–609.

Rocheleau, D., Thomas-Slayter, B. and Wangari, E. (eds) (1996) *Feminist Political Ecology: Global Issues and Local Experiences*, New York: Routledge.

Seager, J. (2009) "Death by degrees: taking a feminist hard look at the 2 degrees climate policy," *Women, Gender and Research* 18(3/4): 11–22.

Smucker, T. A. and Wisner, B. (2008) "Changing household responses to drought in Tharaka, Kenya: vulnerability, persistence and challenge," *Disasters* 32(2): 190–215.

Stockholm Environment Institute (SEI) (2009) "Economics of Climate Change in Kenya." Available online at: www.sei-international.org/mediamanager/documents/Publications/Climate-mitigation-adaptation/kenya-climatechange.pdf (accessed May 19, 2013).

Sultana, F. (2010) "Living in hazardous waterscapes: gendered vulnerabilities and experiences of floods and disasters," *Environmental Hazards* 9: 43–53.

Terry, G. (2009) "No climate justice without gender justice: an overview of the issues," *Gender and Development* 17(1): 5–18.

Tschakert, P. (2007) "Views from the vulnerable: understanding climate and other stressors in the Sahel," *Global Environmental Change* 17: 381–96.

Wangui, E. E. (2004) "Links between land use and gendered division of labor along the Mount Kilimanjaro ecological gradient, Kajiado District, Kenya," Ph.D. dissertation, Michigan State University, East Lansing, MI, USA.

Wangui, E. E. (2008) "Development interventions, changing livelihoods, and the making of female Maasai pastoralists," *Agriculture and Human Values* 25(3): 365–78.

Wangui, E. E. (2012) "Pastoralist livelihood change and the neglected gender dimension: a case study of Loitokitok District, Kenya," in Musyoki, A. and Khayesi, M. (eds) *Environment and Development: Selected Themes from Eastern and Southern Africa*, Gaborone: Bay Publishing.

Wangui, E. E., Smucker, T., Wisner, B., Lovell, E., Mascarenhas, A., Maingi, S., Weiner, D., Munna, A., Sinha, G., Bwenge, C., Meena, H. and Munishi, P. (2012) "Integrated development, risk management and community-based climate change adaptation in a mountain-plains system in northern Tanzania," *Journal of Alpine Research* 100(1). Available online at: http://rga.revues.org/1701# (accessed May 19, 2013).

Ziervogel, G. and Zermoglio, F. (2009) "Climate change scenarios and the development of adaptation strategies in Africa: challenges and opportunities," *Climate Research* 40: 133–46.

10 Gender mapping in post-disaster recovery

Lessons from Sri Lanka's tsunami

Ram Alagan and Seela Aladuwaka

Introduction

Geographic Information Systems (GIS) are increasingly used to analyze gender issues in natural disaster and rehabilitation programs (Gaillard and Maceda 2006; Pincha *et al.* 2007; Meinzen-Dick *et al.* 2012) as gender issues have been recognized as vital policy considerations in such programs (Myers 1994; Paul and Bhuiyan 2004; International Federation of Red Cross and Red Crescent Societies 2010; Lu 2011). In natural disaster situations, women and children are frequently the most impacted, since they do not have proper support systems to gain access to resources, information and services. According to Kottegoda:

> [W]omen, especially if they do not obtain timely warnings about related disaster information, or if their mobility is restricted or otherwise affected by cultural and social limitations, are major casualties in disasters. Gender-biases and stereotypes can complicate and prolong women's recovery, such as when women do not seek or receive timely care for physical and mental trauma.
>
> (Kottedoga 2008: 1)

This increased vulnerability of women is true in many natural disasters. For example, a sizable and disproportionate fraction of the over 6,000 people who died after the 1995 Kobe earthquake in Japan were women, and there was little doubt that social factors, especially gender, had been neglected in the warning process. Likewise, the importance of recognizing gender-specific needs during reconstruction following the Indian Ocean tsunami has also been emphasized by scholars and practitioners (Ariyabandu and Wickramasinghe 2003; Goonesekere 2006; de Silva 2007; Kottegoda 2008).

Recent studies of natural disasters employ mapping technologies such as GIS and remote sensing to support emergency management, hazard mitigation, acquisition of tsunami damage information and recovery programs (ESRI 2006; Merati *et al.* 2010; Şalap *et al.* 2011). Integrating the spatial methodology of Participatory Gender Mapping (PGM) in disaster rehabilitation and recovery is currently limited, yet it has shown to be an innovative approach to investigate gender-specific needs in a variety of contexts. In a review of trends in public participation,

Sieber (2006) finds that projects tend to be guided by grass-roots groups and community-based organizations that use GIS as a tool for capacity building and social change. Using a case study of public-led community advocacy, McLafferty also illustrates the mutually enriching interrelationships between GIS and feminist geography (2002).

The application of PGM in post-disaster rehabilitation programs has the potential to address complex issues such as gender equality, empowerment, access, capacity building and social change. This chapter reports the findings of a study that used PGM techniques to identify post-disaster gender needs following the 2004 tsunami that impacted two coastal communities in the Eastern Districts of Sri Lanka. The discussion is organized into five sections. The second section provides a literature review of work that has been done on gender planning in the context of natural disasters. Given the socially and spatially uneven impact of natural disasters (such as the 2004 Indian Ocean tsunami), gender is an important component of post-tsunami rehabilitation and recovery. The third section discusses the specific rehabilitation and recovery efforts that have been designed to address impacted communities in Sri Lanka. The fourth section outlines the background to the case study area and methodology, with an emphasis on how Participatory Gender Mapping is employed in this research. The fifth section is an analysis of the findings that address gender differences in ways related to the handling of resettlement issues, livelihoods and resources needed to rebuild impacted communities. The conclusion summarizes how participatory mapping has the potential to critically examine gender impacts and needs in post-disaster recovery.

Gender, natural disasters and GIS

Gendered aspects of natural disasters are socially constructed under different geographic, cultural and political-economic conditions, yet governments do not always adequately account for the needs of women and children in a disaster situation. According to Enarson, women tend to experience greater impacts and marginalization in the wake of these disasters (2000). Likewise, social relations, including gender relations, are undermined either by natural or human made conflict (Hyndman 2008). Multiple approaches informed by gender scholars emphasize that distinct characteristics of societies and geographic spaces are inherently linked to generate inequality, marginalization and different power relations for women.

Gender plays a significant role in assigning responsibilities within impacted groups and in determining access to and control of resources among groups. Therefore, gender sensitivity is a valid policy consideration during disasters and throughout the recovery process (World Bank Institute 2009). Although natural disaster brings painful memories, positive attitudes about post-disaster recovery can be seen as an opportunity to channel and leverage investments that upgrade the living standards of underprivileged communities and enhance the livelihood of the most marginalized. In fact, observers often note that in communities that are likely to remain at high risk for future disasters, post-disaster recovery provides considerable opportunity to apply principles of sustainable development and hazard reduction (Yonder *et al.* 2005).

Gender planning is considered to be one of the main practices in social research and sustainable development; however, the use of this approach is limited in natural disaster rehabilitation and recovery. Scholars report that one of the fundamental arguments discussed by vulnerability theorists on disaster studies is that in the disaster situation, gender is typically seen as an insignificant component of the many stages before, during and in the aftermath of the natural disaster (Enarson and Morrow 1998; Bolin *et al.* 1998; Bankoff *et al.* 2004; Blaikie *et al.* 2004). Dasgupta *et al.* (2010) and Momsen (2010) claim that in disasters, women provide economic, health and mental assistance to the family and society in post-disaster recovery activities, and they also participate actively in disaster prevention programs. Although women's groups contribute in many ways, their valuable knowledge and contributions are often neglected or marginalized in decision-making. According to Shrader and Delaney (2000), many implementing agencies have not consciously engaged women in disaster reconstruction because they assumed that women's needs would be addressed in projects targeted to "family wellbeing." Yet, in light of women's pivotal roles at the intersection between household and community recovery, it is essential that government institutions prioritize gender planning as a fundamental social and development policy issue rather than merely creating a symbolic institution for women and children's affairs.

At national and international levels, social scientists have found robust evidence that key social structural factors, like income distribution and access to political power, have a strong influence on people's vulnerability to natural disasters (Blaikie *et al.* 2004; Tobin *et al.* 2006; Hyndman 2008). Researchers report evidence of how diverse social relations have an impact on the effects of disasters and the recovery process. Similarly, feminist scholars have developed the concept of intersectionality, which recognizes the importance of factors like race, class, age, disability and gender in assessing the vulnerability of victims of natural disasters. Gender differentiation remains an important reality in many ways. For example, according to Wisner (1993), women are more vulnerable than men and find it difficult to get loans for rebuilding and re-establishing a viable livelihood. Enarson (2000 and 2006) reports that due to their social class, women are more likely to be victims of disasters because they are disproportionately poor and experience more challenges in relief and recovery efforts. Likewise, studies of gender and disaster, or women in conflict, demonstrate that any given natural or human-made disaster negatively influences communities and brings greater challenges to women and children because of their lack of access to resources, services and information (El-Bushra 2000 and 2004; Afshar and Eade 2004; Byrne 1996; Cockburn 1999; Hedman 2007; Fernando and Hilhorst 2006). In the context of this research, Enarson (1998) argues that in many respects, women's needs and experiences are gender specific yet also influenced by class and ethnicity. These perspectives provide critical insights to problems, processes and mechanisms of household and community recovery and reinforce the need to account for the desires of both men and women in disaster-impacted communities during all phases of disaster management.

The lack of careful planning in disaster rehabilitation could negatively affect the wellbeing of women and, specifically, result in oppression and the violation of their rights and freedoms, as well as putting their comprehensive development and survival at risk. Momsen (2010) states that gendered impacts of natural disasters reflect different positions of women in various societies. Moreover, women generally have less access to resources and less representation at all levels of decision-making, which is particularly critical in emergency situations. Therefore, incorporating gender perspectives in natural disaster-related policies and decision-making processes gives careful consideration to cultural, economic, political and social differences and avoids decisions that are highly patriarchal and technocratic. Using a highly technocratic approach to disaster management has the potential to neglect gendered aspects of disaster rehabilitation. Instead, combining social, economic and political factors with technocratic models could support bottom-up, decentralized and community-based styles of problem solving that explicitly rely on local knowledge, imagination and creativity.

Although it has the potential to address gender matters, one of the major criticisms of GIS is that its focus is normally limited in gender-related issues. In the early 1980s, GIS was employed as a technological expertise model or a scientific approach that offered solutions and answers to complex societal and environmental issues. This approach has generated sequences of constructive social theoretical critiques of GIS: (1) GIS is continuingly disempowering certain community groups while empowering others who have privilege of access due to the cost and complexity of the technologies (Harris and Weiner 1998a and 1998b); (2) there is an inaccessibility of geospatial data (Harris and Weiner 1998b; Pickles 1995; Rundstrom 1995); (3) there is limited representation of local geographic knowledge in data creation, data display and decision-making (Harris and Weiner 1998a and Pickles 1995); and (4) there is often a lack of community participation (Goss 1995; Barndt 1998; Pickles 1995; Harris and Weiner 1998a and 1998b). During this major debate about social theoretical critique, "Initiative 19" (Harris and Weiner 1996) was foundational in changing the use of GIS for societal and environmental studies and decision-making. During the last two decades, the scientific community, academic world and decision-making circles have taken GIS into exciting research applications in order to discuss the most complex issues on society and environment, including gender issues.

Kwan (2002a) highlights the linkage between GIS and gender and especially the potential of GIS to offer spatial perspectives on gender-related research. She articulates the importance of innovative thinking on gendered perceptions of GIS and underlines the current limitations of such thoughts and applications. Kwan further suggests that the critical debate has "unintentionally marginalized the contribution of feminist GIS user/researchers and the potential of feminist perspectives for the development of feminist GIS practices" (Kwan 2002b: 271). Bosak and Schroeder (2005) also underline the important role of GIS in gender research and argue for the possibilities of this tool despite inherent biases with regard to data collection and representation. The authors further stress that if these biases are not addressed at a theoretical or epistemological level, the use of GIS for addressing gender issues in development will be severely limited and will undermine gender roles in decision-making. Schuurman and Pratt (2011) approach GIS as a knowledge system from a feminist perspective. According to the authors, the objective of this perspective was not to accept GIS as truth but to understand how it works as a knowledge system. Instead, the authors illustrate that "understanding GIS as a system of knowledge goes beyond this; understanding how GIS produces truth opens opportunities to produce truth otherwise" (Schuurman and Pratt 2002: 298). Feminist approaches to GIS show how this link could play a key role in critical analyses of representation, data access, local knowledge and empowerment issues, as well as in development and environmental decision-making.

Using PGM methodology, this research aims to promote bottom-up practices in disaster rehabilitation, recovery and decision-making. This approach includes issues of representation, local knowledge integration, empowerment, data ownership and the ability of data providers to respond to group concerns. In addition, PGM recognizes the role that institutional norms and practices play in democratizing available information and the decision-making process. Although GIS has been widely used in policy planning and decision-making, the application of this technology in decision-support systems to analyze gender and disaster issues is still limited. Thus, exploring the role of GIS in gender and disaster could be a vital contribution to the inclusion of gender roles in disaster rehabilitation literatures in particular and the intersection of gender and GIS in general.

Post-tsunami rehabilitation and gender planning in Sri Lanka

The 2004 Indian Ocean tsunami resulted in one of the world's most destructive and complex natural disasters in recent history. Nearly 300,000 people lost their lives; 143,000 were reported missing; nearly a dozen countries were affected; and billions of dollars in international aid provided a glimpse of the enormous impact of this catastrophe (ESRI 2006). Sri Lanka experienced tremendous damage due to the tsunami. According to the Active Learning Network for Accountability and Performance in Humanitarian Action (2007) and Bolton Council of Mosques (2007), the tsunami caused well over US $9.9 billion of economic, infrastructural and human development loss in the Asian impact region. After almost a decade, many affected communities are still confronting a series of challenges to overcome this extreme devastation. The December 24, 2004 tsunami damaged approximately 900 km out of 1,340 km of coastal area. It claimed 35,322 human lives, injured 21,441 and left more than half a million people displaced (Government of Sri Lanka and Development Partners 2005). The Central Bank of Sri Lanka also estimated the damage of the tsunami in Sri Lanka to be US $900 million. This island country has not faced a natural disaster of this magnitude in recent history. For the repair and reconstruction of this immense level of damage, rehabilitation programs required adequate skills, disaster management, data and coordination among decision-making institutions. Unfortunately, rehabilitation and recovery programs did not include adequate efforts that were required to measure the impact of this multifaceted natural disaster situation on vulnerable communities and particularly on women and children. Disaster rehabilitation and recovery continues to be limited and laden with many challenges due to the lack of coordination, collaboration and communication among national government and local community institutions.

Although some rehabilitation and recovery continues in Sri Lanka, studies note that women and children tend to receive little attention because their voices are not heard due to various social, political and patriarchal issues (Ariyabandu and Wickramasinghe 2003; Ariyabandu 2006; de Silva 2007; Kottegoda 2008). The Sri Lankan government and local and international NGOs were involved in the 2004–5 tsunami disaster rehabilitation and recovery efforts, but generally lacked the knowledge to engage affected women groups in these efforts (Ariyabandu 2006). According to de Silva, "the hyperactive scamper to deliver tsunami recovery in Sri Lanka in a short period of time at high speed resulted in decisions that had little or no concern for the participation and representation of affected communities, particularly those disadvantaged" (2007: 5). This mismatch between macro-decisions and local requirements led to discrepancies between post-tsunami recovery interventions and immediate needs during the post-tsunami disaster period.

During the post-tsunami rehabilitation and recovery period, NGOs and governmental institutions designed different protocols that were tailored toward reaching specific tsunami-impacted communities. Due to the political nature of the diverse localities in Sri Lanka, institutions were required to adapt governmental protocols for the different impacted communities. Additionally, ongoing political upheavals made travel and communication in some parts of the Eastern Districts very dangerous. Consequently, institutions involved in rehabilitation and recovery had limited access to and information about communities in this region—partly because of the Liberation Tigers of Tamil Eelam (LTTE) insurgency in 2005 and 2006. Although academic research in Sri Lanka has demonstrated that gender components are vital to policy planning and decision-making (Ariyabandu and Wickramasinghe 2003), the Sri Lankan government lacked such gender awareness and gender-sensitive planning. This deficiency was evident in the aftermath of the tsunami as government planning efforts failed to recognize prevalent gender gaps (Kottegoda 2008). Thus, integrating feminist geography in gender mapping identifies the complexity of gender and other social categories as well as the spatial perspectives of gender needs. Although there were several local and international NGOs engaged in post-disaster rehabilitation and recovery, this approach was not used to identify the needs of impacted communities in Sri Lanka. This study demonstrates that gender mapping can be effective in rehabilitation and recovery by shedding light on, and incorporating the needs of, both men and women.

Background to case study and participatory gender mapping

Two rural communities (Kalmunai and Kinniya) in the Eastern District of Ampara, Sri Lanka, were selected as study areas for this analysis of post-tsunami rehabilitation and recovery (Figure 10.1). The rationale for selecting these two communities was that they were the most devastated after the tsunami in terms of lives lost, property damage, destruction of land and impacted livelihoods. According to the Sri Lanka Census (2011), Muslims are the majority ethnic group in both study areas. In Kinniya, the Muslim population is 58,447 (96 percent), while the Tamil population is 2,522 (4 percent). In Kalmunai, Muslims are 46,682 (64 percent) while Tamils are 26,630 (36 percent). Tsunami-impacted community members and local government representatives participated in the study and assisted the authors in collecting information for investigating post-tsunami recovery plans.

In order to fully appreciate the gendered vulnerability related to the tsunami disaster, it is important to understand the pre-disaster social and economic conditions

Figure 10.1 Post-tsunami disaster rehabilitation and recovery study areas in Sri Lanka. (Fieldwork by Ram Alagan, 2005.)

of this region. Muslims in Kinniya, Trincomalee, largely populate the coastal community of Rahmaniya Nagar (Rahmaniya City). Rahmaniya Nagar is also a traditional community where men mostly engage in fishing and agriculture while women attend to domestic responsibilities. The devastation of the tsunami has nearly destroyed their everyday lives, and local communities have been forced to search for different livelihoods. Many areas of this coastal community were destroyed, and landscapes and shorelines were permanently changed. Kinniya is one of the oldest and most heavily populated settlements in Eastern Sri Lanka and exhibits a deep and rich socio-cultural heritage. The neighboring villages are noted for their remarkably mixed ethnic communities, practices and diverse liveli-hood activities. The long-established Muslim community earns the majority of its income from the deep-sea fishing tradition, while the most Tamils are involved in agricultural activities. The 2004 tsunami absolutely disoriented the community socially, culturally, and economically.

This research was part of a major study on the post-tsunami needs assessment in the World Vision and International Center for Ethnic Studies (ICES). The ICES has been primarily involved in the post-tsunami needs assessment, which indicated the need for the development of an extensive impact report for coastal commu-nities that were damaged due to the tsunami wave. This project was funded by World Vision, in order to implement and support rehabilitation programs in the impact regions. Nearly six months of fieldwork (February 2005 to August 2005) was conducted on post-tsunami needs assessments in several coastal impacted communities in the Eastern Districts. These districts have been characterized by a fragile political state and unstable social conditions among Muslims, Tamils and Sinhalese since the 1983 insurgency. Ruwanpura (2009) also conducted field-work on disaster and development relief in this region and confirms the backdrop of endemic fault lines of war and inequality against which communities nego-tiated the recovery process. Such trends led to ethnic violence among peoples and created anxiety and divides among these coastal communities. Due to the nature and complexity of the local culture, Eastern District post-disaster reha-bilitation and recovery efforts required careful planning and impartiality among the organizations involved. Overall, the lack of understanding of the local cul-ture led to mismanagement, poor coordination of resource utilization and lack of accountability in delivery of disaster recovery and, as a consequence, further increased ethnic tension in the communities.

The study's PGM research design includes a number of participatory methods comprising mental maps, community mapping, ethnographic studies, narratives, photographs and numerous participatory workshops. The use of a mixed-methods design seeks to enhance equitable gender participation, improve gender represen-tation and effectively incorporate local knowledge of women in the outcomes and discussions of the study. PGM also supports initiatives from bottom-up decision-making, rather than top-down or male-dominated, initiatives as a means of identifying the needs of tsunami-impacted communities and, especially, gender disparities in these recovery efforts. In this study, gender mapping tools such as participatory mapping were used with impacted female and male groups separately. Participants described their priorities in post-disaster rehabilitation and recovery planning, and with the facilitation of the research team, maps were drawn with each of the impacted groups.

Fifteen focus group interviews were conducted in both case study areas for women and men using PGM methodology. Male and female participants were trained for the gender mapping exercise and taught the skills for sketching mental maps in order to illustrate their immediate requirements in a post-disaster program. PGM exercises were conducted in temporary camps, where people from the affected communities stayed. Until proper temporary settlement facilities were assembled, the local government authorities directed impacted families to resettle in school buildings as well as built camps. It was emotionally difficult for the research team to encourage the impacted groups to participate in the PGM process, as they were still mentally and physically traumatized from the loss of their properties and loved ones in the tsunami. However, with a great deal of empathy and negotiation, several of the tsunami-displaced community members agreed to participate in the PGM process. Once they began the mapping exercise, participants showed interest and engaged in discussions. Groups of both men and women conducted the activity; most were willing to draw maps and sketches and provide details.

In sum, linking local knowledge, vulnerability and gender needs was fundamental in designing sustainable rehabilitation and recovery efforts in this program. Also, through participatory methods such as focus groups, the research team collected a great deal of information in order to understand the basic needs of women and men. Overall, PGM is an effective way of assessing the strengths and weaknesses of disaster rehabilitation and recovery. As outlined here, understanding multiple representations of post-disaster needs through gender mapping leads to more appropriate and gender-sensitive post-disaster efforts.

Participatory gender mapping and research findings

As described above, top-down technocratic approaches often undermine local voices in post-disaster rehabilitation and recovery programs. Developing sustainable programs in this area requires practitioners to understand community needs and priorities, which are articulated here by local community members through gender mapping. Findings from the Participatory Gender Mapping exercise utilized in this research are presented in three areas that focus on: resettlement, livelihoods and services; needs assessment; and locating resources to rebuild impacted communities.

Mapping resettlement, livelihoods and services in post-tsunami communities

Both the coastal communities of Kalmunai and Kinniya are heavily populated; thus, it is not easy to relocate the impacted groups from the coastal regions. Relocating traditional communities to new locations could also bring negative consequences due to traditional livelihoods and cultural attachments to place.

Conducting PGM is a complicated process because it aims to evaluate the day-to-day lives of people, security, cultural values, religious beliefs and properties rights. During the PGM activity, relocation of community and livelihood needs became an imperative part of the mapping process and created extensive discussion among the participants from both groups. Some participants believed that relocation is important and some did not. These needs were considerably different for men's and women's groups with regard to relocation and livelihood activities. Figures 10.2 and 10.3 illustrate some of the community maps generated from the PGM project that inform our interpretation of women's and men's different priorities, needs and vulnerability in post-tsunami disaster reconstruction and recovery.

Despite the tsunami disaster and devastation, men were willing to resettle in the same vicinity (the coastal region) due to fishing activities, attachment to place, markets, fishing gear, fishing boat safety, cooling facilities and meeting places. Men stated that they were not farmers, and they were unable to cultivate land and participate in agricultural production in inland areas. In addition, they were worried about social and cultural differences among fishing and agrarian families between coastal and inland regions. The socially constructed differences of family values play a major role among these communities in terms of marriage, employment, business and socializing. Although these men's livelihoods and belongings disappeared in the tsunami, it would be difficult for them to alter their culture and livelihood. In the PGM process, men pinpointed the locations of primary resources to rebuild their communities (Figure 10.3). For decades they have resided in these coastal regions and established strong cultural sentiments, livelihoods and societal values that changed significantly after this disaster.

In contrast, as shown in Figure 10.2, women in both communities preferred to relocate inland, since these locations provide more opportunities for them to engage in viable livelihoods. Women from both coastal communities also believed that inland areas are more secure and better for resettling than coastal areas. In general, female participants worried about family members' safety more than employment and properties. A sixty-year-old woman in Kalmunai explained her tragic experience as a result of the tsunami:

> I have five girls and three are in a stage of marriage. I do not work and my family was totally dependent on my husband's fishing employment. I saw him last on Dec 23rd night. He went to sea for employment and he did not come back home. I lost him in the tsunami and I lost our primary home and other two new homes which we built for the dowry of two of my elder daughters. Everything is gone and now I am sitting in the roadside for simple help for foodstuff. My girls and I do not have skills to do any jobs and I am thinking of how I am going to support my girls and their future. Simply, I do not understand what tomorrow means to my family.

Many female participants in Kinniya emphasized that existing settlements near the coastal region should be relocated from Rahmaniya Nagar to Periya Kinniya and Thambalagamam (Figure 10.2). Kinniya is located in heart of the Koddiyar Bay of Sri Lanka's eastern coast, where tsunami waves penetrated nearly 300 m inland and caused severe damage to lives and properties. One of the depressing events of this disaster was the devastation to the general hospital of Kinniya, located within 100 m of the coastal zone. The tsunami wave was 10 m in height and penetrated the whole area. Many lives were lost, including several pregnant mothers. This particular loss is one of the most important reasons given by women

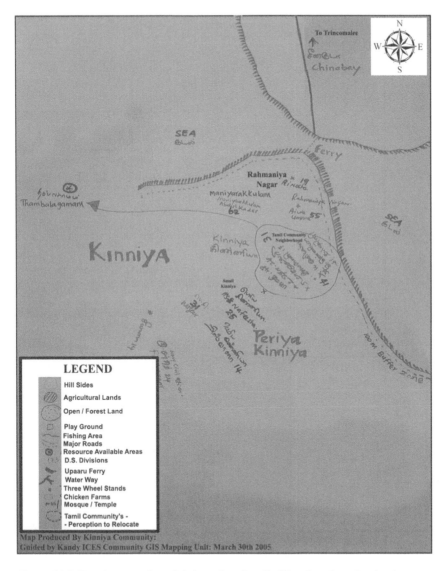

Figure 10.2 Female perception of shelter relocation, livelihoods and service development. (Kandy International Centre for Ethnic Studies (ICES) GIX Unit and Field Survey, Ram Alagan, 2005.)

in the PGM exercise to support their unwillingness to continue their lives in the coastal region.

The personal experiences of the post-disaster devastating scenarios that were incorporated into the PGM activity enabled women and men to articulate their concerns. In disaster, women tend to be more vulnerable than men; hence, they have very specific needs to support their families and most importantly, to safeguard their husbands from danger. The women's groups felt that losing their loved ones

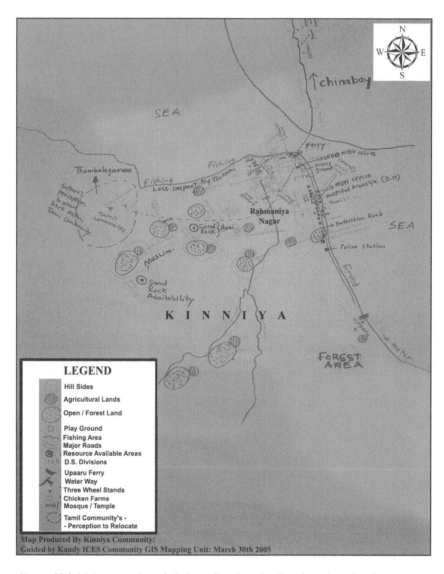

Figure 10.3 Male perception of shelter relocation, livelihoods and service development. (Kandy International Centre for Ethnic Studies (ICES) GIX Unit and Field Survey, Ram Alagan, 2005.)

could not be an option anymore. Moreover, women have no experience in fishing, and thus they find it difficult to earn a living on the coast. Fear of events such as this explains why women in the community mapping exercise located their new settlements further from the coastal region and more to the interior (Figure 10.3). As one middle-aged female participant from Kalmunai stated:

> I have lost all of my properties including home and home gardens. But I am so fortunate to save my family although we suffered severe injuries. Looking at my neighbor's loss, I am so lucky to have my children and husband. I am totally devastated by looking at the disaster to my community. I am not concerned about losing our livelihood or my husband's fishing employment. My prime focus is protecting my husband and children. All others are secondary. I am ready to go inland in search of a new life.

Although they had difficulties sketching these needs through lines, dots and polygons, female participants learned how to situate their priorities in the PGM exercise. However, participants drew a great deal of qualitative information about community needs (in short- and long-term recoveries) and most importantly priorities in rehabilitation and recovery. Gender mapping should include the community in data collection through participatory methods, as well as address the needs and gender priorities of the impacted community. These techniques will help to avoid the sense among participants of being involved in another useless exercise.

Assessing gender needs through participatory mapping

Men and women have varied needs and priorities and are differentially engaged in the rehabilitation and recovery process. These gendered differences became obvious during the PGM exercise among impacted community members. Table 10.1 shows gender disparities, perceptions, differences and requirements in post-tsunami rehabilitation and recovery. Most of the female participants identified immediate priorities as family protection and resettlement sites in the interior. In contrast, men emphasized their priorities as resettlement sites along the coast, job protection and rebuilding marketing facilities in the coastal region. The priorities identified in this exercise underscore gender differences during and after the tsunami disaster, as noted by female and male groups in Kinniya and Kalmunai. Thus rehabilitation is not only about building homes and roads but includes a number of important cultural, economic, political and environmental factors that need to be considered in this process.

During the first phase of the tsunami reconstruction, many institutions tended to focus on physical rehabilitation instead of issues such as gender impact, livelihood impact and human development. Because men and women play different roles in the community, their roles in rehabilitation and recovery vary. These gendered roles were vividly demonstrated in the PGM exercise. The information provided by local residents was a vital source of data for sustainable rehabilitation programs

Table 10.1 Women's and men's rehabilitation and recovery plans in Kinniya, Sri Lanka

Women's priorities	*Men's priorities*
• Family protection and safety	• Resettlement sites at seaside
• Resettlement sites in interior	• Job protection
• Rebuild community integration culturally appropriate settlement	• Rebuild marketing
	• Rebuild sports complex
• Rebuild religious places	• Rebuild transportation facilities
• Sanitary supplies and privacy	• Rebuild schools
• Rebuild schools	• Rebuild fishing facilities
• Opportunities for worship	• Rebuild health facilities
• Rebuild health facilities	• Rebuild youth facilities

Source: field research in Kinniya, 2005.

and illustrated different viewpoints on decision-making and gender-based needs (such as women's emphasis on fear and insecurity in relocating their homes far from the seashore). Therefore, gender-based needs have to be accommodated in disaster rehabilitation, because this makes policy and recovery programs more effective.

Locating resources to rebuild impacted communities

In the immediate aftermath of the tsunami, finding basic resources to rebuild the communities was considered one of the vital tasks. Due to the severe damage to the neighboring physical landscape, finding essential resources and transporting them via the broken road systems were not easy tasks. Additionally, disaster rehabilitation and recovery requires extensive knowledge of infrastructure, construction material, livelihoods, food, clothing, water and sanitation. It is difficult for many women to relate to these aspects of disaster relief, given their cultural and social backgrounds. In the PGM process, women demonstrated limited geographic knowledge of local space, neighborhoods and available resources to rebuild the impacted communities. Knowing the local neighborhood and environment is very important to this reconstruction effort and restoration of livelihood. When the research team asked women to show availability of resources in surrounding areas, most of them had difficulties locating and identifying these places. However, they were very interested in finding resettlement locations.

In contrast, men were able to locate resources such as minerals, water, and forest resources (Figure 10.3). Men also demonstrated their local knowledge and familiarity with the area by indicating in their sketches the location, direction and distance to necessary resources for rehabilitation and recovery. This exercise indicates men's skill at place recognition—a skill that basically stems from gender-based social responsibility and freedom of mobility for men in these communities. Compared to women, mobility among men is higher in this society; thus, their knowledge about neighborhoods and availability of resources in surrounding areas is better. It is clear that the different gender roles that men and women play are linked to gender differences particularly in decision-making and knowledge about local space. Moreover, women's relative lack of knowledge about the surrounding regions, and especially the availability of resources, is due to their limited mobility, since women are less likely to walk and travel outside their home areas.

Conclusion

Women and children were the most affected groups following the 2004 tsunami in Sri Lanka. Although disaster management activities took place in many parts of the island, a number of disaster rehabilitation and recovery programs overlooked or ignored gender needs and disparities, which should be a vital issue for these programs. Research using participatory mapping critically analyzes the spatial context of these communities' loss and the impacts of the disaster where societal structures and governments do not adequately address gender needs. In order to understand the gender needs in rehabilitation and recovery programs, this research employs participatory gender mapping to distinguish needs of men and women. As Enarson (2000) argues, seeing disasters through women's eyes is vital and raises a new set of questions, concerns and issues in order for planners to identify critical gaps and to place gender considerations at the center of development and disaster recovery plans.

In terms of post-disaster rehabilitation and recovery, men are generally more interested in rebuilding their occupations and prefer to stay close to shore, while women are more concerned with the protection of children and families. These research findings show that men emphasize economic livelihoods in post-disaster recovery in comparison to the emphasis that women place on issues that are pivotal in re-building a secure location for their families. GIS has been employed in various social and environmental applications, but it is somewhat underutilized in analyzing gender issues. In contrast, PGM is an effective spatial, as well as community-based, approach to address gender needs and priorities. This approach has the potential to identify gender differences, vulnerability and requirements in post-disaster rehabilitation and recovery programs.

Finally, this study illustrates that PGM has the potential to enhance effective participation of women and men in decision-making in post-disaster situations. The interactive and engaging process inherent in participatory mapping allows meaningful discussions of issues that have a critical bearing on community welfare. This process helps to highlight different needs among various groups in the affected communities, where members are often rendered powerless in decision-making. Through PGM, both women and men are able to raise serious concerns during the practice of drawing sketches, and this provides the opportunity for both decision-makers and impacted communities to engage in better decision-making. PGM also promotes more public participation and negotiation in post-disaster rehabilitation and recovery programs.

The interactive nature of PGM provides opportunities to bridge local knowledge with expert knowledge instead of top-down decision-making, which frequently undermines the very people who are affected by the natural disaster. This approach demonstrates that policy makers should be aware that women have various limitations, which have to be taken into account in sustainable post-disaster rehabilitation and recovery programs. Recognition of these needs is expected to assist disaster management professionals in improving the effectiveness of disaster response and recovery efforts. Given these lessons from Sri-Lanka's tsunami, this research concludes that participatory techniques, such as gender mapping, can help to understand and mitigate the future impact of disasters on families and communities.

References

Active Learning Network for Accountability and Performance in Humanitarian Action (2007) "The boxing day tsunami in numbers stats and facts: scale of the devastation of the boxing-day tsunami." Available online at: www.alnap.org/pool/files/tsunami-stats-facts.pdf (accessed June 2, 2013).

Afshar, H. and Eade, E. (eds) (2004) *Development, Women and War: Feminist Perspectives*, Oxford: Oxfam.

Ariyabandu, M. (2006) "Gender issues in recovery from the December 2004 Indian Ocean tsunami: the case of Sri Lanka," *Earthquake Spectra* 22(S3): 759–75.

Ariyabandu, M. and Wickramasinghe, M. (2003) *Gender Dimensions in Disaster Management: A Guide for South Asia*, Colombo, Sri Lanka: ITDG South Asia.

Bankoff, G., Freks, G. and Hilhorst, D. (eds) (2004) *Mapping Vulnerability: Disasters, Development, and People*, Sterling, VA: Earthscan.

Barndt, M. (1998) "Public participation GIS—barriers to implementation," *Cartography and Geographic Information Systems* 25(2): 105–12.

Blaikie, P., Cannon, T., Davis, I. and Wisner, B. (2004) *At Risk: Natural Hazards, People's Vulnerability and Disasters*, 2nd edition, New York: Routledge.

Bolin, R., Martina, J. and Allison, C. (1998) "Gender inequality, vulnerability and disaster: issues in theory and research," in E. Enarson and B. H. Morrow (eds) *The Gendered Terrain of Disaster: Through Women's Eyes*, Westport, CT: Praeger Publishers, pp. 27–43.

Bolton Council of Mosques (2007) "The Boxing Day Tsunami—Facts and Figures. Relief Work." Available online at: www.thebcom.org/ourwork/reliefwork/96-the-boxing-day-tsunami-facts-and-figures.html?showall=1 (accessed June 2, 2013).

Bosak, K. and Schroeder, K. (2005) "Using geographic information systems for gender and development," *Development in Practice* 15(2): 231–7.

Byrne, B. (1996) "Towards a gendered understanding of conflict," *IDS Bulletin* 27: 331–40.

Cockburn, C. (1999) "Gender armed conflict and political violence," Background paper, The World Bank, Washington, DC. Available online at: http://repository.forcedmigration.org/pdf/?pid=fmo:5013 (accessed June 2, 2013).

Dasgupta, S., Siriner, I. and Sarathi, P. (2010) *Women's Encounter with Disaster*, London: Front Page Publication.

de Silva, A. (2007) "Involuted democracy: tsunami aid delivery and distribution in Ampara District, Sri Lanka," United Nations Development Programme, Bangkok, Thailand. Available online at: http://regionalcentrebangkok.undp.or.th/practices/governance/a2j/docs/CaseStudy-11-SriLanka-InvolutedDemocracy.pdf (accessed June 2, 2013).

El-Bushra, J. (2000) "Transforming conflict: some thoughts on a gendered understanding of conflict processes," in S. Jacobs, R. Jacobson, and J. Marchbank (eds) *States of Conflict: Gender, Violence and Resistance*, London: Zed Books, pp. 66–86.

El-Bushra, J. (2004) "Fused in combat: gender relations and armed," *Development in Practice* 13(2 and 3): 252–65.

Enarson, E. (1998) "Through women's eyes: a gendered research agenda for disaster social science," *Disasters* 22(2): 157–73.

Enarson, E. (2000) "Gender and natural disasters," Recovery and reconstruction department,Geneva. Available online at: www.ilo.int/wcmsp5/groups/public/—ed_emp/—emp_ent/—ifp_crisis/documents/publication/wcms_116391.pdf (accessed June 2, 2013).

Enarson, E. (2006) "Women and girls last?: averting the second post-Katrina disaster," Understanding Katrina: perspectives from the social sciences. Available online at: http://understandingkatrina.ssrc.org/Enarson (accessed June 2, 2013).

Enarson, E. and Morrow, B. H. (eds) (1998) *The Gendered Terrain of Disaster: Through Women's Eyes*, Westport, CT: Praeger Publishers.

ESRI (2006) "GIS and emergency management in Indian Ocean earthquake/tsunami disaster," White paper. Available online at: www.esri.com/library/whitepapers/pdfs/gis-and-emergency-mgmt.pdf (accessed June 2, 2013).

Fernando, U. and Hilhorst, D. (2006) "Every day practices of humanitarian aid: tsunami response in Sri Lanka," *Development in Practice* 16(3): 292–302.

Gaillard, J. C. and Maceda, A. (2006) "Participatory three-dimensional mapping for disaster risk reduction," International Institute for Environment and Development (IIED). Available online at: www.preventionweb.net/english/professional/publications/v.php?id=26929 (accessed June 2, 2013).

Goonesekere, S. (2006) "A Gender analysis of tsunami impact: relief, recovery and reconstruction in some districts in Sri Lanka," CENWOR and UNIFEM, Colombo: Sri Lanka.

Goss, J. (1995) "Geographic information systems and the inevitability of ethical inconsistency," in J. Pickles (ed.) *Ground Truth: The Social Implications of Using Geographic Information Systems*, New York: The Guilford Press, pp. 130–70.

Government of Sri Lanka and Development Partners (2005) "Sri Lanka: post tsunami recovery and reconstruction. Available online at: http://siteresources.worldbank.org/INTTSUNAMI/Resources/srilankareport-dec05.pdf (accessed June 2, 2013).

Harris, T. and Weiner, D. (1996) "GIS and Society: the social implications of how people, space, and environment are represented in GIS," Scientific report for the Initiative 19 specialist meeting, National Center for Geographic Information and Analysis. Available online at: www.ncgia.ucsb.edu/Publications/Tech_Reports/96/96-7.PDF (accessed June 2, 2013).

Harris, T. and Weiner, D. (1998a) "Community integrated GIS for land reform in Mpumalanga Province, South Africa," paper presented at the Empowerment, Marginalization, and Public Participation GIS meeting, Santa Barbara, CA.

Harris, T. and Weiner, D. (1998b) "Empowerment, marginalization, and 'community-integrated' GIS," *Cartography and Geographic Information Systems* 25(2): 67–76.

Hedman, E. E. (ed.) (2007) "Dynamics of conflict and displacement in Papua, Indonesia," Working Paper, No. 42, Refugee studies Centre, University of Oxford, United Kingdom.

Hyndman, J. (2008) "Feminism, conflict and disasters in post-tsunami Sri Lanka," *Gender, Technology and Development* 12(1): 101–21.

International Federation of Red Cross and Red Crescent Societies (2010) "World disaster report 2010: Focus on urban risk." Available online at: www.ifrc.org/Global/Publications/disasters/WDR/wdr2010/WDR2010-full.pdf (accessed June 2, 2013).

Kottegoda, S. (2008) "In the aftermath of the tsunami disaster: gender identities in Sri Lanka." Available online at: www.isiswomen.org (accessed June 2, 2012).

Kwan, M.-P. (2002a) "Feminist visualization: re-envisioning GIS as a method in feminist geographic research," *Annals of the Association of American Geographers* 92(4): 645–61.

Kwan, M.-P. (2002b) "Is GIS for women?: reflection on the critical discourse in the 1990s," *Gender, Place and Culture* 9(3): 271–9.

Lu, A. (2011) "Stress and physical health deterioration in the aftermath of Hurricanes Katrina and Rita," *Sociological Perspectives* 54(2): 229–50.

McLafferty, S. L. (2002) "Mapping women's worlds: knowledge, power and the bounds of GIS," *Gender, Place and Culture* 9(3): 263–9.

Meinzen-Dick, R., Koppen, van B., Behrman, J., Karelina, Z., Akamandisa, V., Hope L. and Wielgosz. B. (2012) "Putting gender on the map: methods for mapping gendered farm management systems in Sub-Saharan Africa," International Food Policy Research Institute. Available online at: www.ifpri.org/sites/default/files/publications/ifpridp01153.pdf (accessed June 2, 2013).

Merati, N., Chamberlin, C., Moore, C., Titov, V. and Vance, C. T. (2010) "Integration of tsunami analysis tools into a GIS workspace—research, modeling, and hazard mitigation efforts within NOAA's center for tsunami research," Geospatial techniques in urban hazard and disaster analysis, NOAA/PMEL/NCTR/JISAO, Seattle, WA. Available online at: http://ecoinfodev.science.oregonstate.edu/files/ecoinfodev/fulltext-3.pdf (accessed June 2, 2013).

Momsen, J. (2010) *Gender and Development*, 2nd edition, London: Routledge.

Myers, M. (1994) "Women and children first: introducing a gender strategy into disaster preparedness," *Focus on Gender* 2(1): 14–16.

Paul, K. and Bhuiyan, H. (2004) "The April 2004 tornado in the North Central Bangladesh: A case for introducing tornado forecasting and warning systems," Quick Response Research Report No. 169, Natural Hazards Center for the University of Colorado. Available online at: www.colorado.edu/hazards/research/qr/qr169/qr169.pdf (accessed June 2, 2013).

Pickles, J. (1995) "Representations in an electronic age," in J. Pickles (ed.) *Ground Truth: The Social Implications of Using Geographic Information Systems*, New York: The Guilford Press, pp. 1–13.

Pincha, C. R., Regis, J. and M. Maheswari (2007) "Understanding gender differential impacts of tsunami and gender mainstreaming strategies in tsunami response in Tamil Nadu, India," *Oxfam America*. Available online at: www.gdnonline.org/resources/Gender_mainstreaming_Pincha_etal.pdf (accessed June 2, 2013).

Rundstrom, R. A. (1995) "GIS, indigenous peoples, and epistemological diversity," *Cartography and Geographic Information Systems* 22(1): 45–57.

Ruwanpura, K. N. (2009) "Putting houses in place: rebuilding communities in post-tsunami Sri Lanka," *Disasters* 33(3): 436–56.

Şalap, S., Ayça, A., Akyürek, Z. and Yalçıner, A. C. (2011) "Tsunami risk analysis and disaster management by using GIS: a case study in southwest Turkey, Göcek Bay Area." Available online at: http://agile.gis.geo.tudresden.de/web/Conference_Paper/CDs/AGILE%202011/contents/pdf/shortpapers/sp_78.pdf (accessed June 3, 2013).

Schuurman, N. and Pratt, G. (2002) "Care of the subject: feminism and critiques of GIS," *Gender, Place and Culture* 9(3): 291–9.

Shrader, E. and Delaney, P. (2000) "Gender and post-disaster reconstruction: the case of hurricane Mitch in Honduras and Nicaragua," World Bank Report, Washington, DC.

Sieber, R. (2006) "Public participation geographic information systems: a literature review and framework," *Annals of the Association of American Geographers* 96(3): 491–507.

Sri Lanka Census (2011) "Population by ethnicity and district according to divisional secretary's division 2012." Available online at: www.statistics.gov.lk/PopHouSat/CPH2011/index.php?fileName=pop32&gp=Activities&tpl=3 (accessed June 2, 2013).

Tobin G. A., Bell, H. M., Whiteford, L. M. and Montz, B. E. (2006) "Vulnerability of displaced persons: relocation park residents in the wake of Hurricane Charley," *International Journal of Mass Emergencies and Disasters* 24(1): 77–109.

Wisner, B. (1993) "Disaster vulnerability: scale, power, and daily life," *Geojournal* 30(2): 127–40.

World Bank Institute (2009) "Why gender issues in recovery are important?", International Recovery Platform. Available online at: www.recoveryplatform.org/assets/tools_ guidelines/Why%20gender.pdf (accessed June 3, 2013).

Yonder, A., Akcar, S. and Gopalan, P. (2005) "Women's participation in disaster relief and recovery," The Population Council, New York. Available online at: www.popcouncil. org/pdfs/seeds/Seeds22.pdf (accessed June 2, 2013).

11 Ecodevelopment, gender and empowerment

Perspectives from India's Protected Area communities

Ruchi Badola, Monica V. Ogra and Shivani C. Barthwal

Introduction

Over 150,000 protected areas (PAs) covering at least 24 million km^2 (IUCN/UNEP 2009) exist for the purpose of safeguarding terrestrial and marine-based flora and fauna, thereby helping to maintain essential ecosystems and to conserve the Earth's unique natural biological heritage. While the ecological benefits of biodiversity conservation accrue at local, national, regional and global scales, the social costs of conservation have been borne disproportionately by members of local communities. Over 50 percent of PAs worldwide are inhabited by local populations (Torri 2011) and in India, this figure is at least 65 percent (Kothari *et al.* 1989). Negative local impacts associated with the creation of PAs typically include loss of livelihoods, conflicts with wildlife and park authorities, forcible relocation and social and cultural displacement; indigenous and poor communities are especially vulnerable to these negative impacts (Kothari *et al.* 1996; Brechin *et al.* 2003; West *et al.* 2006).

Gender-based divisions of labor in these communities typically assign the responsibility for the collection of PA-based resources that fulfill domestic needs (such as fuelwood, fodder, water and edible or medicinal plants) to women. However, women are often structurally excluded from institutions of environmental management due to the persistence of traditional gender-based power hierarchies that privilege men's knowledge and experiences (Guijt and Shah 1998), or which include them only as tokens (Agarwal 2001). Such hierarchies intersect with other economic, ethnic and cultural structures of discrimination and bias (Rocheleau *et al.* 1996; Elmhirst and Resurreccion 2008). In rural India especially, gender and class/caste-based hierarchies tend to collectively reinforce longstanding patterns of elite and male privilege, authority, and knowledge about matters environmental or otherwise (Agarwal 1992 and 2000; Badola 1998).

As this chapter will explore, these long-established trends may be changing in PA communities that experiment with Integrated Conservation and Development

Projects (ICDPs), known in India as "ecodevelopment." Numerous studies from around the world have shown that caste/class issues strongly shape local-level experiences with both PAs and ecodevelopment (see, for example, Kothari *et al.* 1996; Rangarajan and Saberwal 2003; McShane and Wells 2004; Woodroffe *et al.* 2005; Baviskar 2003; Saberwal and Chhatre 2003). More specific analyses of the ways in which gender functions as an equally critical and mediating variable are rare and scattered in the literature on PAs in India (for exceptions, see case studies by Chandola *et al.* 2007; Ogra 2008; Pandey 2008; Torri 2010). However, Flintan (2003) and Vernooy (2006) are good examples of existing work on gender and natural resource management in sub-Saharan Africa and other parts of Asia, respectively. This chapter attempts to help redress the gender gap in the literature on ICDPs by specifically focusing on the gendered nature of contributions to ecodevelopment project participants around Indian PAs. In so doing, it also furthers understanding about the conditions under which ecodevelopment can be a means for empowerment for both women specifically, and their communities.[1]

For both instrumentalist and ethical reasons, PA managers and conservation advocates in India have begun taking a more active interest in linking "conservation" with vaguely defined notions of "women's empowerment" as part of their overall approach to ecodevelopment (Alers *et al.* 2007; Mishra *et al.* 2009; Ogra 2012a). The promotion and creation of women's collective groups (sometimes a variant or extension of a commonly maintained, traditional women's institution known as *mahila mangal dal*) remains a key feature of this approach (Pillai and Suchintha 2006; Rao 2006). Promotion of livelihood diversification strategies reflecting active engagement with markets, microfinance and home-based activities (along with the creation of new spaces for women to discuss and prioritize village-level social and environmental issues) have also been key features of these groups (Ogra 2012a). While an emerging literature examines the short- and long-term implications of microfinance for gender relations (see Aladuwaka and Oberhauser, this volume) and feminist political ecology continues to examine the links between collective action and empowerment outcomes (e.g. Rocheleau *et al.* 1996; Agarwal 2000; Resurreccion and Elmhirst 2008; Cruz-Torrez and McElwee 2012; Parpart *et al.* 2002; Cornwall and Anyidoho 2010; Ebyen 2011), relatively little is known about what the conservation community's deepening interest in women truly suggests for gender relations in sites targeted for ecodevelopment or in terms of the advancement of women's practical and strategic needs (Moser 1993) at individual or collective scales.

This chapter seeks to foster a much-needed discussion of these important issues within both the conservation community and among feminist researchers. We contribute to this wider discussion by presenting relevant field-based perspectives from four PAs in the Indian Himalayas. Our guiding questions are as follows: To what extent are contributions to ecodevelopment planning and related outcomes linked to gender? What, if any, have been the gender-differentiated impacts of ecodevelopment? And what lessons can be drawn for improving the design, effectiveness and empowerment potential of the ecodevelopment model in practice? To begin, the following section provides a contextual overview of the ecodevelopment experience in India.

Ecodevelopment in India: stages of theory and practice

The passage of the Indian Wildlife Protection Act in 1972 led to rapid expansion of a PA network that presently covers approximately 4.74 percent of the country (WII 2012). From just six national parks and fifty-nine wildlife sanctuaries in 1970 (WII 2012), the total number of PAs at the close of 2012 was 664 (IUCN/UNDP 2009). However, this growth has been accompanied by an intensification of both discursive and physical conflicts about the meaning and practice of conservation (Gadgil and Guha 1992; Torri 2010). Nevertheless, while retaining the assumption that inviolate core zones are required for long-term and effective conservation planning, PA policies and practices around the world slowly began to embrace the "participatory" approaches used in the field of rural development throughout the 1980s and 1990s.

In 1982, ecodevelopment was proposed by a task force of the Indian Board for Wildlife as a new strategy to reduce people–park conflict (IWBL 1993). Envisioned as a site-specific set of new incentives for conservation, including support for rural development and alternative income-generating opportunities to reduce forest dependence, and to and fulfill the promise of community participation, the primary objective of ecodevelopment in this first stage (1982–92) was to compensate local communities for lost access to resources and to reduce their dependence on the PA (Badola 1995). In practice, these initiatives tended to be isolated and fragmented "development" activities, without clear and direct linkages to conservation through sustainable resource use. Projects centered on infrastructure development as compensation for the curtailed use of PA resources (e.g. construction of schools, water tanks, community halls or village approach roads), leaving PA authorities "grappling with explanations to justify these as 'ecodevelopment' " (Mishra *et al.* 2010: 1362). At some sites, ecodevelopment interventions included the introduction of new varieties of hybrid cattle, distribution of smokeless *chullahs* (stoves), training in beekeeping and experiments with ecotourism (Karlsson 1999)—each intended to induce a reduction in biomass extractions from PAs.

An intensified emphasis on local participation (which tightened conservation-development linkages), partnerships with local stakeholders and conservation NGOs (intended to generate public support for the nation's PAs), and a focus on creation of village-based microplans characterized a second stage of ecodevelopment. Chief among the eighty government-supported initiatives in place by the mid 1990s, was the India Ecodevelopment Project (IEP) (1996–2001), described as "perhaps the most widely debated wildlife project ever undertaken in India" (Singh and Sharma 2004: 300). Funded by the World Bank/GEF, the IEP covered seven PAs and included some of the most important tiger reserves. Central to the IEP and related approaches was the creation of a new village-based institution: the ecodevelopment committee (EDC). With a member-secretary from the

Forest Department to maintain the conservation interests of the PA, chief among the EDC's duties was to create a site-specific set of ecodevelopment planning objectives and activities through means of participatory processes (Bhardwaj and Badola 2007) and awareness building about the value of conservation, the PA network and sustainable resource use (Singh and Sharma 2004). Typical EDC microplans included PA-based activities such as: habitat restoration and improved protection measures; village-based activities such as growing of useful fuel/fodder species at the PA border; distribution of alternative cooking technologies; and receiving development assistance (Karlsson 1999; Baviskar 2003; Pandey 2008; Mishra *et al.* 2010).

Ecodevelopment in the 1990s continued to be premised on the assumption that a direct relationship exists between poverty alleviation of PA communities and improved PA protection (Mishra *et al.* 2009). However, given that the *sine qua non* of ecodevelopment continues to be conservation of PA resources (and not development of livelihoods in the PA communities), the overall approach has drawn much criticism. Observers have argued that this form of development has inadequate participatory aspects (Singh and Sharma 2004; Pandey 2008), is prone to corruption (Karlsson 1999; Pandey 2008), serves the interests of the elite/powerful (Baviskar 2003; Dejouhanet 2010) and excludes women from meaningful involvement (Chandola *et al.* 2007; Ogra and Badola 2008). While they have emerged primarily in response to practices in the context of the IEP, these critiques have also led us to question whether a possible third stage of practice is emergent, i.e. one in which empowerment is implicitly (or explicitly) among the objectives. The following section addresses our collective field-based experiences in the Indian Himalayas for insight and comparative examples.

Ecodevelopment in the Indian Himalayas

The tremendous historical, cultural and economic diversity characteristic of India's PA network complicates attempts to generalize. Thus, we focus on the cultural landscapes of the Indian Himalayas, which are similar enough to provide useful points of comparison, yet are also sufficiently varied to support analysis and discussion of the transformative potential of ecodevelopment practice. With approximately179 PAs (WII 2012) for the protection of the ecologically fragile landscape and biodiversity conservation, the Indian Himalayan region (Figure 11.1) is also highly significant from a social standpoint. The high dependence on natural resources, consolidation of the PA network and subsequent denial of traditional resource rights, have led to widespread mistrust, alienation and loss of livelihood security. Women living in villages adjacent to PAs have been strongly affected by the designation of PAs due to the gender-based division of labor typical to the region.

Ecodevelopment projects in and around four Himalayan PAs were initiated with the objectives of improving biodiversity conservation outcomes though reduction of local-use pressures, minimizing conflicts and improving the well-being of the local people. While community empowerment was envisioned

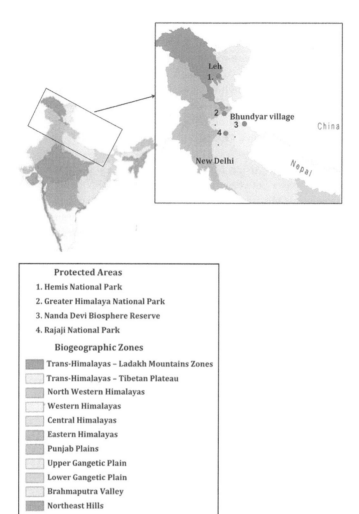

Figure 11.1 Selected protected areas and biogeographic zones in India. (Field survey, 2006.)

through participation-related objectives, some sites had specific goals emphasizing women's empowerment. However, conceptualizations of empowerment varied between sites (see Tables 11.1–3). Our combined fieldwork at these sites spans the period 1997–2012. During this period, at least one of us has supervised or conducted a wide range of data collection activities in the case study communities. The research methods have been largely qualitative, emphasizing participant-observation, in-depth interviews, focus group discussions, household questionnaire surveys and literature review. Participatory rural appraisal (PRA) exercises were also conducted in each PA community; for example, participatory mapping exercises were used to identify gender-based uses of space, and gender-based time budget analysis was used to help us identify ecodevelopment-related labor practices. Participant-observation and interviews were used to gather personal narratives about participants' experiences with ecodevelopment. The following section contains our findings from the four case studies.

Hemis National Park

The Hemis National Park (HNP), located near the city of Leh (Jammu and Kashmir State) (Figure 11.1), is home to wild carnivores of significant conservation value (Table 11.1). Human population density is significantly lower at 3.5 persons/km^2 than that for the state of Jammu and Kashmir (99.7 person/km^2) in the approximately twenty-three hamlets inside the HNP. Local communities in and around HNP are relatively homogeneous in terms of religion, culture and traditions (Table 11.1). While outmigration of males aged fifteen to forty years is common, traditional practices of animal husbandry (rearing goats, sheep, yaks and pack animals) continue to be one of the main occupations of people living inside the HNP boundary. The average household agriculture landholding size is 2.05 acres and the average annual household income is approximately US$1,853 (Chandola 2012). Though differences in economic status are observable, our fieldwork suggests that these differences have not resulted in the high societal stratification typical of the rest of India. The gender dynamic is liberal; cultural practices of polyandry help to explain why women have traditionally been accorded high status in Ladakhi society and are central to household-level decision-making (Rizvi 1983; Norberg-Hodge 1991).

Since the creation of the HNP in 1981, the denial of basic infrastructure facilities (such as roads) has resulted in negative local attitudes toward the PA management. Medical and market facilities are reached by foot after a half- to two-day journey, which has been seen by locals as a lost opportunity for employment. In an effort to respond to these grievances and to further reduce anthropogenic pressure on the park, ecodevelopment activities were initiated in 2003. Several types of tourism associated with income-generating activities were introduced, including trekking home-stays, "parachute" cafés and camping. At the time of our most recent fieldwork (2011), nearly all the households along the trek routes of the HNP had set up guest rooms for tourists as part of the initiative. The J&K Wildlife Protection Department involved the local NGO, Snow Leopard Conservacy (SLC), in order to develop a strategy to improve the sustainability of local communities' income in consultation with the community members (Jackson *et al.* 2003) (Table 11.2).

During in-depth fieldwork (2005–6 and 2011), we observed that men and women both participate actively in these activities and that the sphere of labor-related activities for males and females is spatially segregated along gendered lines. While men work as guides, "pony men" and "trek operators" on routes in the PA, women are mainly responsible for home-based activities including managing guests and making local products for sale, such as apricot jam and woollens

Table 11.1 Socio-economic and geographic characteristics of selected Protected Areas in India

	Hemis National Park (HNP)	Great Himalaya National Park (GHNP)	Nanda Devi Biosphere Reserve (NDBR)	Rajaji National Park (RNP)
Protected Area category[a]	National park	National park	Biosphere reserve	National park
Area[a]	4,750 km^2	1,171 km^2	5,881 km^2	826 km^2
Biodiversity focus[a]	Snow leopard, mountain ungulates	Snow leopard, Asiatic black bear, Himalayan brown bear	Snow leopard, leopard, Himalayan black and brown bears, Himalayan musk deer	Large mammals (tiger and elephant)
Location/ biogeographic zone[a]	Trans Himalayas	North Western Himalayas	Western Himalayas	Upper Gangetic Plain
Local community	Ladakhi Buddhist[b]	Himachali[e,f]	Garhwali and Bhotia (Tolcha)[h]	Garhwali, Gujjars[k]
Employment/alternatives/ opportunities	Animal husbandry, Tourism, self-employment[b]	Agriculture, horticulture, animal husbandry[e,f]	Agriculture, animal husbandry, hydro-electric projects, tourism, services, NTFP extraction[h]	Agriculture, animal husbandry, services, handicrafts, NTFP extraction[k]
Religious composition of the community	Buddhist[b]	Hindu and Muslim[e,f]	Hindu[h]	Hindu and Muslim[k]
Community composition in terms of class	Homogenous[b]	Heterogeneous	Heterogeneous	Heterogeneous
Typical gender dynamics in terms of gender differentiated roles	Liberal[e]	Conservative[e,f]	Some areas are liberal and some areas are conservative[h]	Conservative[k,l]
Human–wildlife conflict	Prominent (58% households suffer livestock loss)[b,d]	Prominent (A total of 1,322 livestock loss during 1989–1998)[g]	Prominent (98% households reported crop damage due to wildlife)[i]	Prominent (91% of study households reported crop loss due to wildlife)[l]
People–park relation pre-ecodevelopment	Mistrust: not ready to negotiate[b]	Hostile: aggressive to the PA staff and property[e,f]	Hostile: aggressive to the PA staff[i]	Hostile: aggressive to the PA staff[k,l]
People–park relation post-ecodevelopment	Friendly[d]	Friendly[e,f]	Conflict on how to manage the resources[j]	Moving towards neutral[k,l]

Sources: [a]WII (2012); [b]Jackson and Wangchuk (2004); [c]Norberg-Hodge (1991); [d]Chandola (2012); [e]Saberwal and Chattre (2003); [f]Baviskar (2003); [g]Chauhan (1999); [h]Silori and Badola (1995); [i]Bosak (2008); [j]Sabic (2011); [k]Chandola et al. (2007); [l]Ogra, 2008.

Table 11.2 Overview of ecodevelopment initiatives in selected Protected Areas in India

	Hemis National Park	Great Himalayan National Park	Nanda Devi Biosphere Reserve	Rajaji National Park
Year of ED* initiative	1999 (started as APPA – Appreciative Participatory Planning and Action)[a]	1994[c]	1999 (Bhundyar Valley EDC)	1999[g]
Stages of community participation in ED	Socially acceptable contributing significantly to people's livelihood[a]	Socially acceptable contributing significantly to people's livelihood[d]	Socially acceptable contributing to alternative livelihood[e,f]	Have not moved much beyond micro-planning and awareness
Role of NGO in ED	Facilitator[a]	Provider of marketing links[d]	None	None
Programs under ED	Local home-stays, treks and cafés' (livelihood activities)[a]	Village-stay; treks; capacity building; Women Saving and Credit Groups (WSCG)[d]	Waste management; capacity building; eco-fee collection[e,f]	None
Committee managing ED initiatives	Amma Chokspa (mother group, i.e. women groups), Youth society[b]	Devta (diety) Committees; EDC; WSCG[c,d]	EDC; Women's Self-help Groups[e,f]	EDC; Women's Self-help Groups[g,h]
Committee goals	Employment; self-sustenance[a,b]	Employment; self-sustenance[d]	Employment; self-sustenance[e,f]	None
Funds utilized for	Plantations, civil works such as maintenance of footpaths, monastery etc.; capacity building	Civil works such as maintenance of footpaths, temples etc.; capacity building[c,d]	Civil works such as maintenance of footpaths, temples, etc; capacity building; waste management[e,f]	Civil works such as maintenance of footpaths, temples, community halls, etc.[g,h]

Sources: [a]Jackson and Wangchuk (2004); [b]Author's observation; [c]Saberwal and Chattre (2003); [d]Pandey (2008); [e]Seaba (2006); [f]UNEP and WCMC (n.d.); [g]Chandola et al. (2007); [h]Ogra and Badola (2008).

Note:
*Ecodevelopment (ED)

Table 11.3 Impacts of ecodevelopment in selected Protected Areas in India

	Hemis National Park	Greater Himalaya National Park	Nanda Devi Biosphere Reserve	Rajaji National Park
Gender sensitive	Formulated opportunities for men and women on the basis of their gender roles[a]	At later stage of ED,* gender role recognized[c,d]	No recognition of gendered roles[e]	No recognition of gendered roles[g,h]
People's reaction to ED	Embrace[a]	Embrace[c,d]	Embrace[e,f]	Neutral[g,h]
Changes in social relations post-ED	Provided equal opportunity to all[a,b]	Intensified upper and lower class differences[c,d]	Economic up-liftment of those who were directly involved in ED activities[e,f]	None[g,h]
Employment opportunity for women	Yes[a,b]	Not specifically until WSCG formed[c,d]	Not specific[e,f]	Not specific[g,h]
Changes in gender relations	Men and women empowered through economic and social opportunities[a,b]	WSCGs empowered women through economic, social and political opportunities; men inspired by the success of WSCGs[c,d]	Men empowered through economic and social opportunities; younger, educated women empowered through economic opportunity[e,f]	None[g,h]
Conservation outcome	Conservation-conscious community[a,b]	Conservation-conscious community[c,d]	Conservation-conscious community[e,f]	None[g,h]

Sources: [a]Jackson and Wangchuk (2004); [b]Chandola (2012); [c]Saberwal and Chattre (2003); [d]Pandey (2008); [e]Seaba (2006); [f]UNEP and WCMC (nd); [g]Chandola et al. (2007); [h]Ogra and Badola (2008).

Note:
*Ecodevelopment (ED)

(Figure 11.2). Women also manage the "parachute cafés" (freestanding tea and snack bars constructed from discarded parachute cloth). The cafés are located near the villages, enabling women to balance their cafe-related and household responsibilities. While managing home-stay is an individual household responsibility, managing parachute cafés and camping sites are communal activities which ensure equal participation and benefit-sharing among the participating households (Figure 11.3). Additional revenue is generated from fees collected for use of campsites near the village grazing grounds (approximately US$1 per tent). Part of the total revenue generated from camping sites and parachute cafés goes to a common village fund and/or the *gompa* (monastery) fund while the rest is distributed among the families managing the camping sites and the café. The decision on revenue sharing is through mutual agreement among the villagers (men and women). The egalitarian gender relations in Ladakhi society have made it possible for both men and women to participate equally in decision-making pertaining to village and monetary issues at the household level and even at the village forums (Table 11.3).

The creation of homestay-based tourism has made it possible for women to participate in income-generating activities in new ways (contributing to overall

Figure 11.2 Ladakhi woman picking apricots. (Photo by Shivani C. Barthwal, 2006.)

Figure 11.3 Nimaling pasture at Hemis National Park. (Photo by Shivani C. Barthwal, 2006.)

goals of "community" empowerment as well as to individual-level empowerment), since the money from such activities mostly goes directly to the women. However, women in participating households from which working-age members have out-migrated have relatively greater labor demands. This scenario may prove to be less of a disadvantage than appears at first sight. During our fieldwork, for example, we were assigned one such home for the stay, as it was this particular household's turn. Realizing that our stay was to be solely organized and run by a woman who had a toddler in her lap, we were concerned for her: How would she manage both her guests as well as her baby? Her response was that "such good income" (US$11 per guest, per night) was well worth the modest efforts required to provide clean beds and food to the visitors; in addition, she expressed the opinion that the interesting social interactions with her guests was an additional benefit. During the times when she needed to work intensively for the guests, it was observed that the neighbors happily offered to look after her child. We found that this favor was reciprocated and was a common occurrence.

Great Himalayan National Park

The Great Himalayan National Park (GHNP) and adjacent conservation areas are located in Kullu district of Himachal Pradesh and together encompass an area of

1,171 km². Of this area, 754.4 km² are designated as the GHNP, while 265.6 km² are contained in an ecodevelopment-focused "ecozone" that lies adjacent to two Wildlife Sanctuaries (Sainj WLS and Trithan WLS). On the western edge of the park, approximately 160 villages are located in a 5 km wide ecozone. The local community is comprised predominantly of agro-pastoralists, who follow a religious tradition of local deity (*devta*) worship (Table 11.1). Villages are stratified into deeply entrenched caste-based categories in which the more powerful *Brahmins* and *Rajputs* occupy one end of a spectrum, and poorer and historically "untouchable" communities (known as Scheduled Castes or SCs) lie at the other. Members of *Brahmin* and *Rajput* communities are better endowed in terms of land and access to resources than SCs (Tucker 1997). As with caste-mediated interactions, a relatively conservative gender dynamic (privileging men) can be observed; this dynamic places a disproportionate and large burden of farm, forest and household labor demands squarely on women.

The global criticism of exclusionary practices and increasing local resentment for lost livelihood opportunities led to inclusion of the GHNP in the IEP. Throughout the project, the initiatives were focused on creating and increasing membership of EDCs, irrespective of membership criteria and linked to a need to have representatives from all sections of society. The resulting EDC was thus dominated by village elites, who were also the members of the powerful *Devta* committees (Saberwal and Chhatre 2003). While a conservative gender dynamic contributed to an atmosphere of non-participation by upper-caste women among both Hindus and Muslims, caste hierarchies among the Hindus formed a barrier for the meaningful participation of members of the SCs. The EDC was eventually declared unsuccessful and funding was stopped in 1999 (World Bank 2002) (Table 11.2). Engagement with ecodevelopment nevertheless continued under a park director who sought to directly integrate women's interests into new ecodevelopment activities (Pandey 2008). Women of poor and PA-dependent households, who had little or no participation in prior village development activities, were targeted for organization into Women Saving and Credit Groups (WSCG) of twelve to fifteen members each. Members were encouraged to save at least one rupee (conversion rate: INR1 = US$55) daily and accept the group's interest-free loans in support of new and "alternative" livelihood activities (Table 11.3).

Our site visits, observations and interactions with WCSG members and ecozone residents over the past ten years demonstrate that communities in these areas are now successfully operating small-scale businesses using their own savings. They are generating new income through activities such as vermi-composting, medicinal herb propagation, apricot oil production, hemp-based handicrafts and the cultivation of organic vegetables and cash crops. Moreover, benefits to the entire family accruing from women's contributions to the household income base are leading to changes in women's status and household standing. WSCG members engaged in personal capacity building through literacy classes and training for value-addition of the local produce are finding enthusiastic support from their male household counterparts. Some of the WSCG members have since been elected to governance positions as members of local administrative bodies. Encouraged by the changes associated with this "pro-women" approach to ecodevelopment, many men have sought participation as well; for example, men have undergone training for new work as tourism guides, porters, cooks, and, perhaps most importantly from a conservation perspective, have agreed to give up illegal herb collection from the Park in exchange for the benefits of ecodevelopment. Park-supported strengthening of market linkages and marketing efforts have also helped to provide the critical opportunity for locals to sell their products within the wider state of Himachal Pradesh, to retail outlets in the capital of New Delhi, and even abroad (United Kingdom).

Nanda Devi Biosphere Reserve

Nanda Devi Biosphere Reserve (NDBR) comprises three zones: buffer, transition, and two core areas (the Nanda Devi National Park and Valley of Flowers National Park) (Table 11.1). There are forty-seven villages in the buffer and thirty-three villages in the transition zone. *Bhotia* and *Garhwalis* are the main ethnic groups of the area. Historically, local livelihoods for both communities have been based on trade and marginal agro-pastoralism (Table 11.1). The income of people living in the buffer zone is lower (INR11,100/hh/yr) than that of people living outside the NDBR (INR 13,340/hh/yr) (Saxena *et al.* 2011). While women have traditionally held a high status in both Garhwali and Bhotiya culture (Dash 2006), daily responsibilities and expectations in NDBR follow gender-based divisions. Both communities privilege men in terms of control of money, while women have greater control over household resource allocation in day-to-day living.

In response to the growing impacts of tourism and adverse livelihood consequences that followed the declaration of the Valley of Flowers National Park, ecodevelopment initiatives at NDBR were started in 1999 in Bhundyar village (Seaba 2006) (Table 11.2). By 2003, a range of income-generating options was implemented, such as the collection of "eco-fees" from mule owners and the creation of numerous tourism-related activities (Figure 11.4). Jobs were envisaged to benefit villagers across lines of age and gender: village youths (boys and girls) are paid to collect eco-fees and check receipts, while adult males staff various check-posts, check receipts and operate a mule rotation system. Adult women were encouraged to join the EDC in order to help manage the total revenue collected.

Ecodevelopment in Bhundyar has yielded mixed results in terms of broadly defined community participation, as well as in terms of social empowerment and changes in the overall gender dynamic (Table 11.3). While it has streamlined and regulated tourism in the valley, ecodevelopment has led to employment for only a small number of people—mostly men. In describing shortcomings, Seaba (2006) noted that the distance between villages and the ECD meeting place was an obstacle to widespread member participation. We found this to be particularly problematic for older residents, who were not involved in the day-to-day activities despite being EDC members. Second, while the absence of older women and men was particularly noticeable in the EDC activities, female members in general

Figure 11.4 Entry gate at Bhundyar Valley, Nanda Devi Biosphere Reserve. (Photo by Vinay Bhargava, 2006.)

appeared to have little influence in the ecodevelopment decision-making process. Third, although a few educated young women did get employment opportunities in EDC activities, adult women often acted as proxies for their husbands or silent spectators, and they often limited their participation to simply adding their names to member attendance rosters. Our observations are also consistent with those reported by Silori (2007), who notes that the ecodevelopment activities in the area are characterized by unequal distribution of the economic benefits and lack of employment. Lastly, in contrast to the experiences in HNP and GHNP, we found that there was an overall lack of men's willingness to make real space for women in the ecodevelopment process. On balance, these activities do not appear to have led to significant outcomes for women; rather than presenting an alternative vision for equity through conservation, the creation of the EDC appears to have reproduced existing gender and age hierarchies.

Rajaji National Park

Rajaji National Park (RNP) lies at the foothills of the Himalayas in the state of Uttarakhand. RNP is an important site that is under considerable developmental pressure; it includes a large area of the fragile Shiwaliks system and houses the northwestern-most population of the Asiatic elephant (*Elephas maximus*) (Table 11.1). In addition to a relatively small group of seasonal grass collectors and forest workers, RNP has two main resident communities: *Van Gujjars*, a Muslim pastoral community that relies on park resources to support its buffaloes, and small-scale subsistence agriculturalists. Agricultural communities in and around the park vary widely in terms of degree of religious and ethnic heterogeneity and in terms of caste-based social stratification (Table 11.1). In general, these

societies maintain a conservative gender dynamic in which men dominate transactions related to the local cash economy and maintain control over household
decision-making and resources.

RNP represents one of the more complex examples of people–PA relations due
to the long standing climate of hostility and mistrust between various stakeholders
and user groups (Table 11.1). Resource extraction by the user groups, economic
loss due to livestock predation and crop damage and resulting antagonistic reac-
tions have led to a seemingly intractable situation of conflict. From 1999 to 2002
however, ecodevelopment was initiated at the RNP border in a few PA-dependent
villages on an experimental basis (Table 11.2).

Unlike the cases described above, ecodevelopment work at RNP did not pro-
ceed much further than the microplanning stage. Based on repeated visits to the
sites in the year following the project conclusion, we found little or no evidence
of the empowerment of the community as a result of the ecodevelopment process (Table 11.3). Women were not involved in EDC decisions regarding forests,
resource use and conservation; they were also poorly represented in the committees themselves. In addition, women had very little knowledge or understanding
about the project in general. In a follow-up survey we conducted at one site, for
example, only half of the women respondents even knew about the committee,
and they were ignorant of both its function and of the procedures for becoming a member. Social norms restricted women's participation, especially in study
villages with *pardha* (female seclusion) practice. Conservative gender dynamics
typical in the RNP villages reinforced longstanding behavioral expectations that
women should not speak loudly in front of elders or men—further undermining
the potential for ecodevelopment to promote the active participation of women.

Multiple field visits (2002–7) also showed that women are too busy fulfill-
ing household needs to find time for political participation, particularly when the
meetings are inconveniently scheduled. Years after the project was over, we found
that even those women who recall the EDC initiatives still expressed a feeling
that they were not educated enough to say or contribute anything significant. Men
similarly reported to us a belief that due to women's illiteracy and general levels
of ignorance about "extra" household matters, there was no need for them to participate in any such future meetings. Not surprisingly, perhaps, women reported
that in the future they would prefer to have separate meetings (i.e. without men).
However, women who were otherwise willing to participate expressed reservations
about the mixed-community (user-group) format of the ECD planning meetings.
Those from a higher caste were reluctant to go to a common forum in which
women from lower-caste groups were also invited.

When we returned to the RNP sites in 2012 for another update and to gauge
interest in restarting ecodevelopment activities, we found that the interests of male
and female members continued to reflect their "traditional" gender roles. Asked
how funds should be used if ecodevelopment were to be attempted again, men in
our focus group wanted the funds of EDC to be utilized for providing training and
loans for business, whereas women were eager to reduce forest-based work and
wanted to see afforestation in the village using useful fodder species. In response

to the notion of introducing new cooking gas (LPG) connections, the women we interviewed were more interested to learn about existing government programs intended to promote "development" of poor households, more broadly conceived, than to learn about the gas program specifically.

Participation, gender and ecodevelopment practice in Indian Protected Areas

The case studies in this research offer several insights about the potential for ecodevelopment to intersect in meaningful ways with empowerment objectives, specifically through the support of sustainable livelihoods (DFID 2000) and the reduction of gender-based inequities. We summarize some of the key outcomes of the projects in Table 11.3 and return to our guiding questions to reflect on their meanings below.

India is a country of diverse cultures and class structures, which are represented in the case studies of four National Parks from the Himalayas. As illustrated in Tables 11.2 and 11.3, ecodevelopment activities within these sites varied in their basic approach to involve local communities in conservation initiatives. For example, the activities started at the HNP were targeted at improving the financial capital of the local community by using existing natural and social capital in a sustainable way. However, work initiated in the GHNP and RNP emphasized an objective of creating EDCs (Table 11.2). Unlike in the HNP, activities in these two sites were also funding-driven rather than need-driven. Considerations related to gender issues also varied between sites. In viewing the approach adopted for ecodevelopment in the HNP, it is clear that gender-segregated workplaces were considered as assets rather than obstacles to creating opportunities for meaningful participation of men and women (Table 11.3). However, in the GHNP the need for a gender-sensitive approach came much later and was somewhat in response to disappointment from men about EDC (Table 11.3). At the GHNP, men were ultimately inspired by women's successes. The relatively homogenous nature of Ladakhi society can also be viewed as an asset in the HNP case. Perceived homogeneity helped ecodevelopment to succeed in the HNP because participants felt unified, and the examples of GHNP and RNP illustrate the value of starting with the creation of smaller, homogenous sub-groups prior to scaling up to a necessarily heterogeneous village-level EDC.

Though we can cautiously conclude from our case studies that ecodevelopment is capable of providing meaningful opportunities for empowerment at both collective and individual scales, we would emphasize that simply modifying the EDC institutional structure is not enough to adequately challenge entrenched gender norms and interacting caste/class hierarchies. Rather, we suggest that a wider and transformation-oriented approach is more appropriate; one that enables both men and women of all sub-groups to witness and experience the benefits of participation. Understanding how and when this will be possible will require deep analyses of how livelihoods and power sharing in PA communities are not only gendered and structured by other markers of status, but also how these communities are themselves affected by the broader political economy of the surrounding region. It is within the larger region that their ecodevelopment partners in government, markets and the NGO sector operate, after all—and within which the PAs are themselves situated (driving, for example, male outmigration and other livelihood strategies).

At this slightly broader scale, then, we also suggest that there remains a need to build more trust directly between community and PA representatives as part of changing norms for women's (especially poor and non-dominant caste women's) and other disadvantaged groups' participation in natural resource management, specifically. In order to overcome the cultural constraints that continue to be associated with mixed-gender interactions in this context, women foresters and NGO-affiliated motivators (who otherwise tend to be exclusively male) could be called upon to help further demonstrate the value of engaging women in the process of ecodevelopment planning. However, we would caution that use of female staff alone will not be sufficient and that related institutional efforts to adequately train, retain and support a gender-sensitive staff must continue.

What, if anything, can we conclude about gender-differentiated impacts of ecodevelopment and related questions about the potential for this form of development to support empowerment objectives? We believe that both within and between cultural communities, the ecodevelopment initiatives in the HNP and GHNP have contributed to women's and men's empowerment in meaningful ways. The employment opportunities created through ecodevelopment initiatives in the HNP and GHNP have diversified the livelihood options of the agro-pastoral community and provided them with greater financial security. The income from additional livelihood sources is being used for the education of the younger generation. Moreover, many women in these sites are now responsible for their own monetary decisions and have asserted a greater voice in conservation-related decision-making. In these areas, women's increased agency as actors at both household- and village-levels is being achieved through financial means rather than through the forest-based and "domestic" contributions that mountain women of an earlier generation relied upon for status (Badola and Hussain 2003; Ogra 2008). Women involved in ecodevelopment have also increasingly become champions of wider social causes and developments for their villages/regions. Similarly, men in some sites, having realized the potential of women to contribute to income and the economic security of the household, have joined their wives in their activities or are openly supportive of what they now see as ambitions that benefit the whole family. As one woman in the GHNP told us, "Ecodevelopment has increased the love between husband and wife."

The ecodevelopment initiatives in the RNP and NDBR, on the other hand, have not been able to spur changes in gender relations or promote women's empowerment. This is mainly due to unchallenged male domination of the process and related attempts by elites to capture the lion's share of potential economic benefits. This is a continuation of well-documented patterns of resistance to women's empowerment through participation in rural development projects more

broadly (Gujit and Shah 1998). One possible reason for this unchallenged domination could be the tacit acceptance of "patriarchal bargains" (Kandiyoti 1998)
by both ecodevelopment advocates as well as the women they hope will become
"empowered" through meaningful participation. For example, as a recent study of
community-based wildlife conservation practices ironically illustrates, the success
of a given conservation project may be linked to its ability to work within, rather
than to challenge, traditional gender roles and related gender-based power hierarchies (Ogra 2012b). At the same time, opening a discussion at the RNP about
ecodevelopment did help to reveal the need for improved levels of awareness
within the community about existing and complementary government programs
that address the development needs of poor and female-headed households (often
the same individuals).

We are encouraged by cases where gender-sensitivity through ecodevelopment
has led to the meaningful participation of women, increased livelihood security for both men and women and enhanced nature conservation. However, we
must note that while individual women's practical needs for increased income
and livelihood security are gradually being addressed by livelihood diversification and access to microcredit, it is perhaps still too early to tell whether
women's larger, collective strategic needs (e.g. for equity in terms of power in
other arenas of decision-making and benefit-sharing) are really being addressed
by ecodevelopment initiatives around PAs. As our case studies illustrate, such
outcomes will be largely place-specific and dependent upon the attitudes toward
change by a range of actors and stakeholders. Anticipating the kinds of changes
that may be associated with the introduction of ecodevelopment at any site will
also be predicated upon an understanding of what the PA, itself, has meant for
local communities—and how these meanings, too, are informed by gender and
class/caste.

How can gender analysis ultimately contribute to debates about the meaning
and effectiveness of ecodevelopment? In our view, an underlying cause of the failures of ecodevelopment (and indeed, of community-based conservation in a larger
sense) has been the inability or unwillingness to look beyond "community" and
thoroughly explore its components. Resource managers and conservation advocates have long recognized that class/caste-based societal divisions and hierarchies
play crucial roles in shaping human use of the environment; why is the same not
yet true for gender relations? If advocates of ecodevelopment wish to address
environment and development concerns, they must take into account how the
underlying conflicts are shaped by both more obvious socio-economic factors as
well as less "visible" gender-based dimensions.

It will be important for ecodevelopment advocates to remember, however, that
men and women cannot be essentialized into single homogenous groups and that
gender roles are guided in different ways by place-specific cultural norms; thus,
the meaning and implications of gender equity/inequity changes geographically.
Moreover, gender identities are "always cross-cut by and inscribed in other forms
of inequality" (Kandiyoti 1998: 140), i.e. along lines of caste, class, ethnicity,
religion, age, sexual orientation and other social markers. Nevertheless, it will be

worthwhile for practitioners to, at a minimum, expect to: (1) hold separate meetings for men and women at various stages of EDC creation; and (2) hold them at times and locations that are convenient for community members (rather than the practitioners). As we have argued elsewhere (Ogra 2012b, Badola *et al.* 2012; see also the 2012 Bhutan +10 Declaration), we repeat here our conviction that data collected during and as a result of such meetings should be disaggregated by gender (and ideally subjected to additional, finer scales of analysis). Without even such a basic level of gender analysis to guide planning and practice, ecodevelopment will never reach its potential to truly be a cornerstone of effective and sustainable development practice for PA communities in the future.

Finally, the gender-based examination at the NDBR in particular contributes another layer of complexity by clearly signaling the need to extend analyses into age-based heterogeneity. Ecodevelopment advocates would do well to further consider the implications of inter-generational differences in setting priorities related to PA ecozones and for tourism zones in particular. India's youth represents a vibrant, rapidly growing and increasingly educated sector of the population, with skills and interests that will continue to be valuable in our globalized "information age"—particularly in terms of shaping the direction of sustainable ecotourism around PAs. Their voices must also be heard for any long-term ecodevelopment planning to be viable.

Conclusion

Several overarching trends emerge from this analysis of gender aspects of ecodevelopment within protected areas. First, a commitment by PA authorities to demonstrate the benefits of a gender-centered approach contributed to the positive outcomes observed at the HNP and GHNP. In contrast, ecodevelopment activities at the NDBR and RNP did not focus on the heterogeneity of the community in a meaningful way—and despite rhetoric of participation, an entrenchment of male/elite power perpetuated a longstanding reluctance to share power, especially with women. Second, domination of the ecodevelopment planning process by elites and men at these sites ensured that women developed no real stake in ecodevelopment planning outcomes. Therefore, the limited efforts to challenge or re-envision traditional gender roles (in particular, the division of labor and role in household decision-making) doused any spark of meaningful change that might have been ignited by the planning process. Third, as suggested by the RNP and GHNP cases, a reluctance to cross caste boundaries reminds us that although caste-based discrimination in India has been outlawed for decades, structural inequalities persist as part of the cultural and socio-economic landscape. Fourth, tangible and positive outcomes can serve as models for other potential ecodevelopment sites given communities' willingness to overcome perceived barriers and deep-seated prejudices (against both park authorities and within their own villages).

Although the particular combination of related activities is site-specific, broader communication between and within PA communities about the empowerment potential of ecodevelopment will be critical. Our experience in the Himalayas is necessarily limited to a few PAs across a landscape rich with potential sites for transformative ecodevelopment initiatives. What would a complete gender analysis of such efforts across the entire region—or even the globe—reveal? We close with a call for urgently needed support for a wide range of critical and rigorous studies (both quantitative and qualitative) that can be used to further document the gendered dimensions of the full range of ecodevelopment initiatives, as well as for research that empirically assesses the linked conservation outcomes. Such studies would, by necessity, need to embrace complexity—both of the myriad human systems driving the broader political economy of the PA regions and of the overlapping ecological systems for which the PAs themselves exist. (In many cases, we should note, the human and environmental systems are likely to be at odds with one another.) Additional areas that we feel are currently ripe for study include gender-based analyses of the increased use of microcredit as a path to alternative livelihoods around PAs, detailed examinations of the gendered aspects of PA-based tourism (ecotourism home-stays and guided treks, in particular), focused analyses of gendered social networks and gendered uses of social capital to promote sustainable livelihoods through ecodevelopment and interrogations of the competing uses of "empowerment" in these practices. We look forward to a tide of feminist research that seeks to answer these and other such questions.

Meanwhile, this limited study has shown that weakening gender and class/caste hierarchies through ecodevelopment is in some cases possible. Related efforts to redress the underlying power inequities—a prerequisite for empowerment—will hold an important key in unlocking the transformative potential of sustainable and alternative livelihoods development around PAs, both in India and elsewhere. In addition, although women within and across divisions of caste/class/age/ethnicity will obviously play a major role in this process, we believe that men will play perhaps an even more critical role in terms of responding to increasing demonstrations that improvements for women yield benefits for the entire family and, in turn, their communities. Likewise, PA communities will undoubtedly begin to bear witness to the benefits that follow when park dependence and illegal extraction activities are replaced by self-prioritized, culturally appropriate and low-impact resource use habits that are compatible with prevailing conservation objectives and laws. We are hopeful that the new generation of PA managers and other conservation advocates will be trained in theories and models of sustainable development that emphasize the important role of gender equity and that employ notions of gender that emphasize intersectionalities.

In sum, though ecodevelopment is by no means a panacea, when done in a way that directly includes objectives to reduce and ultimately eliminate gender-based inequities as both the means and the ends, ecodevelopment can, and will, support a broader and deeply meaningful transformative process in and around the world's protected areas. It is a tall order, but one we feel is worth making.

Acknowledgements

The authors wish to acknowledge our home institutions and colleagues for technical, financial and logistical support. We extend special thanks to Alyssa Bosold and Paul DiSalvo for their assistance in manuscript preparation. We also wish to express our deep gratitude to members of the participating ecodevelopment communities and the PA management/staff who continue to support our related research efforts. Finally, we sincerely thank this volume's editors and the two anonymous reviewers for their constructive and helpful comments on earlier drafts of this chapter.

Note

1 While acknowledging the contested and context-specific uses of the term "empowerment," we join the Pathways of Women's Empowerment research consortium in recognizing empowerment as process-based and as occurring for people when they are "able to imagine their world differently and to realise that vision by changing the relations of power that have kept them in poverty, restricted their voice and deprived them of their autonomy" (Ebyen 2011: 2).

References

Agarwal, B. (1992) "The gender and environment debate: lessons from India," *Feminist Studies* 18(1): 119–59.

Agarwal, B. (2000) "Conceptualising environmental collective action: why gender matters," *Cambridge Journal of Economics* 24: 283–310.

Agarwal, B. (2001) "Participatory exclusions, community forestry, and gender: an analysis for South Asia and a conceptual framework," *World Development* 29(10): 1623–48.

Alers, M., Bovarnick, A., Boyle, T., Mackinnon, K. and Sobrevila, C. (2007) *Reducing Threats to Protected Areas: Lessons from the Field*, Washington, DC: World Bank.

Badola, R. (1995) "Critique of people-oriented conservation approaches in India", IDS Study Report, Sussex, England: Institute for Development Studies.

Badola, R. (1998) "Attitudes of local people towards conservation and alternatives to forest resources: a case study from the lower Himalayas", *Biodiversity and Conservation* 7(10): 1245–59.

Badola, R. and Hussain, S. A. (2003) "Conflict in paradise: women and protected areas in the Indian Himalayas," *Mountain Research and Development* 23(3): 234–7.

Badola, R., Barthwal, S. C. and Hussain, S. A. (2012) "India beyond tradition: emerging trends in women's role in protected area management in the Indian Himalaya", paper presented at the Bhutan +10: Gender and Sustainable Mountain Development in a Changing World Conference, Thimphu, Bhutan, October 15–19, 2012.

Baviskar, A. (2003) "States, communities and conservation: the practice of ecodevelopment in the Great Himalayan National Park", in V. K. Saberwal and M. Rangarajan (eds) *Battles over Nature: Science and the Politics of Conservation*, New Delhi: Permanent Black.

Bhardwaj, A. K. and Badola, R. (2007) "An assessment of ecodevelopment initiatives in Periyar Tiger Reserve", Study Report, Dehradun, India: Wildlife Institute of India.

Bhutan +10 Conference Declaration (2012) "Gender and sustainable mountain development in a changing world: international conference," Bhutan +10 Conference, October 15–19, 2012, Thimphu, Bhutan. Available online at: www.icimod.org/bhutan +10/ (accessed May 20, 2013).

Bosak, K. (2008) "Nature, conflict and biodiversity conservation in the Nanda Devi Biosphere Reserve," *Conservation and Society* 6(3): 211–24.

Brechin, S. R., Wilshusen, P. R., Fortwangler, C. L. and West, P. C. (eds) (2003) *Contested Nature: Promoting International Biodiversity Conservation with Social Justice in the Twenty-first Century*, Albany, NY: State of New York Press.

Chandola, S. (2012) "An assessment of human-wildlife interaction in the Indus Valley, Ladakh, Trans-Himalaya," PhD thesis, Rajkot, Gujarat, India: Saurashtra University.

Chandola, S., Badola, R. and Hussain, S. A. (2007) "Factors affecting women's participation in conservation programmes in and around Rajaji National Park, India," *Journal of the Indian Anthropological Society* 42: 111.

Chauhan, N. P. S. (1999) "Evaluation of crop damage in the ecodevelopment project area to suggest mitigation measures," FREE-GHNP 04/18. Report prepared for the Wildlife Institute of India and Winrock International, Dehradun, India: Wildlife Institute of India.

Cornwall, A. and Anyidoho, N. A. (2010) "Women's empowerment: contentions and contestations," *Development* 53(2): 144–9.

Cruz-Torres, M. L. and McElwee, P. (2012) *Gender and Sustainability: Lessons from Asia and Latin America*, Tucson, AZ: University of Arizona Press.

Dash, C. (2006) *Social Ecology and Demographic Structure of Bhotias: Narratives & Discourses*, New Delhi, India: Concept Publishing Company.

Dejouhanet, L. (2010) "Participatory ecodevelopment in question: the case of the Parambikulam wildlife sanctuary in South India," *Alpine Geography Review (Revue de Geographie Alpine)* 98(1): 83–96.

Department for International Development (DFID) (2000) *Sustainable Livelihoods Guidance Sheets*. Available online at: www.livelihoods.org/info/info_guidancesheets.html (accessed May 20, 2013).

Elmhirst, R. and Resurreccion, B. (2008) "Gender, environment, and natural resource management: new dimensions, new debates," in B. Resurreccion and R. Elmhirst (eds) *Gender and Natural Resource Management: Livelihoods, Mobility, and Interventions*, London: Earthscan.

Eyben, R. (2011) "Supporting pathways of women's empowerment: a brief guide for international development," Pathways Policy Paper, Brighton, UK: Pathways of Women's Empowerment Research Programme Consortium. Available online at: www.pathwaysofempowerment.org/Supporting_Pathways_of_Womens_Empowerment. pdf (accessed May 20, 2013)

Flintan, F. (2003) *Engendering Eden*, London: International Institute for Environment and Development (IIED).

Gadgil, M. and Guha, R. (1992) *This Fissured Land: An Ecological History of India*, Berkeley, CA: University of California Press.

Guijt, I. and Shah, M. K. (eds) (1998) *The Myth of Community: Gender Issues in Participatory Development*, New Delhi: Vistaar Publications.

Indian Board for Wildlife (IWBL) (1983) *Eliciting Public Support for Wildlife Conservation*, Report of the Task Force, New Delhi: Government of India.

IUCN/UNEP (2009) "The World Database on Protected Areas" *(WDPA)*, Cambridge: UNEP-WCMC. Available online at: www.wdpa.org/AnnualRelease.aspx (accessed May 20, 2013).

Jackson, R. and Wangchuk, R. (2004) "A community-based approach to mitigating livestock depradation by snow leopards," *Human Dimensions of Wildlife* 9(4): 307–315.

Jackson, R., Mishra, C., McCarthy, T. and Ale, S. B. (2003) "Snow leopards?: conflict and conservation," *Biology and Conservation of Wild Felids* 1:417–430.

Kandiyoti, D. (1988) "Bargaining with patriarchy," *Gender Studies* 2(3): 274–90.

Kandiyoti, D. (1998) "Gender, power, and contestation: rethinking bargaining with patriarchy," in C. Jackson and R. Pearson (eds) *Feminist Visions of Development: Gender Analysis and Policy*, New York: Routledge.

Karlsson, B. J. (1999) "Ecodevelopment in practice: Buxa Tiger Reserve and forest people," *Economic and Political Weekly* 34(30): 2087–94.

Kothari, A., Pandey, P., Singh, S. and Variava, D. (1989) "Management of national parks and sanctuaries in India," status report, New Delhi: Indian Institute of Public Administration (IIPA).

Kothari, A., Singh, N., and Suri, S. (eds) (1996) *People and Protected Areas: Towards Participatory Conservation in India*, New Delhi: Sage Publications.

McShane, T. O. and Wells, M. P. (eds) (2004) *Getting Biodiversity Projects to Work: Towards More Effective Conservation and Development*, New York: Columbia University Press.

Mishra, B. K., Badola, R. and Bhardwaj, A. K. (2009) "Social issues and concerns in biodiversity conservation: experiences from wildlife protected areas in India," *Tropical Ecology* 50(1): 147–61.

Mishra, B. K., Badola, R. and Bhardwaj, A. K. (2010) "Changing dimensions of biodiversity conservation with stakeholders participation in India—paths ahead," *Indian Forester* 1361–71.

Moser, C. O. (1993) *Gender Planning and Development: Theory, Practice, and Training*, New York: Routledge.

Norberg-Hodge, H. (1991) *Ancient Futures: Learning from Ladakh*, San Francisco: Sierra Club.

Ogra, M. (2008) "Human–wildlife conflict and gender in protected area borderlands: a case study of costs, perceptions, and vulnerabilities from Uttarakhand (Uttaranchal), India," *Geoforum* 39(3): 1408–22.

Ogra, M. (2012a) "Gender mainstreaming in community-oriented wildlife conservation: experiences from nongovernmental conservation organizations in India," *Society and Natural Resources* 25: 1258–1276.

Ogra, M. (2012b) "Gender and community-oriented wildlife conservation: views from project supervisors in India," *Environment, Development, and Sustainability* 14(3): 407–24.

Ogra, M. and Badola, R. (2008) "Compensating human-wildlife conflict in protected area communities: ground-level perspectives from Uttarakhand, India," *Human Ecology* 36(5): 717–29.

Pandey, S. (2008) "Linking ecodevelopment and biodiversity conservation at the Great Himalayan National Park, India: lessons learned," *Biodiversity and Conservation* 17(7): 1543–71.

Parpart, J., Rai, R. and Staudt, K. (2002) *Rethinking Empowerment: Gender and Development in a Global/Local World*, New York: Routledge.

Pillai, K. R. and Suchintha, B. (2006) "Women empowerment for biodiversity conservation through self-help groups: a case from Periyar Tiger Reserve, Kerala, India," *International Journal of Agricultural Resources, Governance and Ecology* 5(4): 338–55.

Rangarajan, M. and Saberwal, V. K. (eds) (2003) *Battles over Nature: Science and the Politics of Conservation*, Delhi: Permanent Black.

Rao, A. (2006) "Making institutions work for women," *Development* 49(1): 63–7.

Resurreccion, B. and Elmhirst, R. (eds) (2008) *Gender and Natural Resource Management: Livelihoods, Mobility, and Interventions*, London: Earthscan.

Rizvi, J. (1983) *Ladakh: Crossroads to High Asia*, New Delhi: Oxford University Press.

Rocheleau, D., Thomas-Slayter, B. and Wangari, E. (eds) (1996) *Feminist Political Ecology: Global Issues and Local Experience*, New York: Routledge.

Saberwal, V. K. and Chhatre, A. (2003) "The Parvati and the Tragopan: conservation and development in the Great Himalayan National Park," *Himalayan Research Bulletin* 21(2): 79–88.

Sabic, K. (2011) "Human-wildlife conflicts in the Nanda Devi Biosphere Reserve, Uttarakhand, India," unpublished Bachelor of Arts thesis, University of Michigan (Ann Arbor).

Saxena, K.G., Liang, L. and Xue, X. (2011) *Global Change, Biodiversity and Livelihoods in Cold Desert Region of Asia*, Dehradun, India: Bishen Singh Mahendra Pal Singh.

Seaba, N. (2006) "Public participation: rhetoric or reality? An analysis of planning and management in the Nanda Devi Biosphere Reserve," unpublished Master of Natural Resources Management thesis, University of Manitoba: Natural Resources Institute.

Silori, C. S. (2007) "Perception of local people towards conservation of forest resources in Nanda Devi Biosphere Reserve, north-western Himalaya, India," *Biodiversity Conservation* 16: 211–22.

Silori, C. S. and Badola, R. (1995) *Nanda Devi Biosphere Reserve: a study on Socioeconomic Aspects for the Sustainable Development of Resource Dependent Population*, First Study Report, Dehradun, India: Wildlife Institute of India.

Singh, S. and Sharma, A. (2004) "Ecodevelopment in India," in T. O. McShane and M. P. Wells (eds) *Getting Biodiversity Projects to Work: Towards More Effective Conservation and Development*, New York: Columbia University Press.

Torri, M. C. (2010) "Decentralising governance of natural resources in India: lessons from the case study of Thanagazi Block, Alwar, Rajasthan, India," *Law, Environment & Development Journal* 6(2): 230–45.

Torri, M. C. (2011) "Conservation approaches and development of local communities in India: debates, challenges and future perspectives," *International Journal of Environmental Sciences* 1(5): 871–83.

Tucker, R. (1997) *The Historical Development of Human Impacts on Great Himalayan National Park*, FREE-GHNP 04/14, report prepared for the Wildlife Institute of India and Winrock International, Dehradun, India: Wildlife Institute of India.

Vernooy, R. (ed.) (2006) *Social and Gender Analysis in Natural Resource Management: Learning Studies and Lessons from Asia*, New Delhi: Sage Publications.

UNEP and WCMC (n.d.) "Nanda Devi and Valley of Flowers National Parks Uttaranchal, India," World Heritage Sites, Protected Areas and World Heritage.

West, P., Igoe, J. and Brockington, D. (2006) "Parks and peoples: the social impact of protected areas," *Annual Review of Anthropology* 35(1): 251–77.

Wildlife Institute of India (WII) (2012) National wildlife database, Dehradun, India: Wildlife Institute of India.

Woodroffe, R., Thirgood, S. and Rabinowitz, A. (eds) (2005) *People and Wildlife: Conflict or Coexistence*, New York: Cambridge University Press.

World Bank (2002) *Implementation Completion Report for the Forestry Research Education and Extension Project*, Washington, DC: The World Bank.

Index

abolitionist 72, 76–9
affective investment 60
African heritage 141
African National Congress (ANC) 151, 158
Agarwal, B. 2, 200–1
agriculture 38–9, 93, 128, 188, 205–6
Aitken, S. 87, 89, 95–6, 100
Aladuwaka, S. 39, 41
Alexander, J. and Mohanty, C. 2, 89, 97, 99, 143
alternative economic strategies 92–4
anti-sex trafficking xvi; legislation 74; policy 9, 71–2, 76, 78, 80; UN Protocol 8, 9
Appalachia 88–91, 96–7; Region 87, 91–2, 99; Appalachian Regional Commission 91
Ariyabandu, M. 186
austerity 59–60

Baba Dogo 25
Badola, R. 200, 202–3, 216, 218
beneficiaries 27, 30, 105, 135
Beneria, L. 2–5, 12, 17, 20, 31
Beni ngoma 146
Bibi Titi Mohamed 10, 139–40, 143, 146, 152
biogeographic zones 204
borrowers 37, 39; small-scale 35; credit 36, 38, 41–2, 44–5; women 43, 45, 48
non-borrowers 36, 41–4, 46, 49
Bob, U. 104, 111
bottom-up 184–5; decision-making 188; approach 48; structure 48

care: care deficit 56, 58; care work 56; caring labor 56
Caribbean: family ties 57; migration 57, 69
Chipko Movement 2

climate: change xvi, 5, 6, 10, 11, 163–5, 168–71, 174, 176–7; adaptation 8, 163, 168, 174; models 164, 176; narratives 167–8; variability 165, 167, 171, 176
Coalition Against Trafficking of Women (CATW) 76, 81
collective economic strategies 87, 101
colonial rule 138, 141
commercialization 18; of public toilets 8, 22, 31; of public services 17, 31
community maps 190
cooperatives 93–4, 97, 99
counter-topographies 88–9, 94, 98–9
credit programs 36, 38, 41–3, 47–9
cross-boundary 88, 94
cross-subsidization 25
cultural norms 44, 92, 217

Dar es Salaam xii, 142, 145, 153–4, 158
decision-making 46–7, 49, 105–6, 109, 119, 122, 124, 135, 183–6, 195; top-down 11, 123
Demeritt, D. 164, 176
development 2, 50, 53, 55, 58, 61–2, 66, 119, 125, 134, 143, 156, 177, 203; studies xv, 36–7, 49, 123, 126; gender and 2; uneven 1, 89, 91–2; feminism and 2–3; globalization and 3–4, 6–7, 9, 11, 17; sustainable 10, 134, 183, 218–19; UN Millennium Development Goals 18; human development index 129–30
diaspora 53–5, 62–4, 66–7; formation 57, 67; Jamaican 58, 62, 65
disaster recovery 8, 188, 195 (see post-disaster recovery)
disembodied workers 58
Diversified Revenue Model 25

domestic: service 57; work 10, 45, 57, 68, 125

ecodevelopment xii, 200–3, 205–8, 211–22; committee (EDC) 202–3, 208–9, 211–15; projects 201–3; planning 201, 203, 216, 218
economic production 61
economic strategies 4, 6, 35, 36, 37, 49, 87–90, 92–5, 98, 101; women's 36, 87
Ecotact Ltd. 23–4, 31
ecotourism 202, 218–19
emotions 54, 56–8, 63, 67, 189; spaces 54; support 56; labor 56; geographies 58
empowerment 4, 6, 35, 45, 49, 99, 108, 185, 210, 215, 219, 220; women's 2, 10, 41, 43, 48, 50, 122, 155, 201, 204, 216
entrepreneurs 31, 49, 52, 54, 61, 67; entrepreneurship 24, 26, 30, 32, 35, 44; activities 62, 69
ethnography 68
event history calendar 165, 176–7

family xviii, 17, 20, 25, 35, 42–4, 56–7, 59–60, 64, 68–9, 105, 112, 126, 128, 130–1, 133, 152, 157, 158, 183, 190, 193–4, 211, 216, 219
female: migration 71, 74; sex-trafficking 72, 74, 76, 80; prostitution 77–8; headed household 106, 110, 217; participants 10, 94, 110, 112, 114, 189–90, 193
feminism 2; transnational xv, 2–6, 9, 87, 89, 97, 99–100; radical 77
feminist: analyses 3, 11, 21, 49; fieldwork 7, 9, 100; geography 3, 35–6, 123, 156, 186; movement 149; political ecology 6, 10–11, 201; scholarship 2, 4, 7, 31; studies 1, 7, 9; theories 1–2
firewood 125–6, 133, 171, 174
focus group 10, 41, 103, 106, 110–11, 165, 189, 204, 214
free choice prostitution 77, 79
freedom of mobility 194

gardens 94, 97, 112, 114, 193
Geiger, S. 142–4, 146, 148–9, 151–2
gender: analysis 23, 26, 28, 217–19; and development 2, 134; asymmetries 63; differences 8, 19, 27, 112, 182, 193, 195; equity 5–7, 11, 17, 217, 219; inequalities 5, 9, 50, 119; planning

182–3, 185; relations 2, 4, 7, 17, 36, 50, 57, 63, 94, 103, 172–3, 176, 182, 201, 208–9, 216–17; responsibilities 163, 172; roles 2, 10, 42, 99, 170, 185, 194, 208, 214, 217–18; studies 2; -segregated workplaces 215
Geographic Information Systems (GIS) 123–4, 132–5, 139, 156, 181–5; participatory 106, 108, 112, 114, 122, 124, 126, 139; mapping 123, 150
geospatial technology 123–5
Ghana 142–3
Gibson-Graham, J. K. 3–5, 94
Global Alliance Against Trafficking of Women (GAATW) 77
global care circuits 57
Global Positioning Systems (GPS) 5, 122, 125–7, 129, 132–5
globalization 1–11, 18, 20, 71, 73, 80, 87–9, 91, 93
grass-roots organizations 5, 6, 11
Great Himalayan National Park (GHNP) 206, 210–11, 213, 215–18
Green Belt Movement 2

Hemis National Park (HNP) 205–8, 210, 213, 215–16, 218
Himalayas 201, 203, 206–8, 210, 213, 215, 219
HIV/AIDS 128, 130
Human Development Index 129–31
human trafficking 80–1

Ikotoilets 22, 24, 28, 30
Inanda 103–4, 106–19
India 2, 11, 62, 98, 205, 215, 218–19; Board for Wildlife 202
Indian Ocean tsunami 11, 181–2, 185
inequality 1, 2, 4, 5, 89, 91–4, 100, 105, 139, 143, 145, 149, 182, 188, 217
informal settlement 22, 113–14, 124
information technology 10, 122–3, 134
infrastructure 18–20, 22, 53, 92, 109, 111–12, 119, 194, 202, 205
inheritance 104–5, 111–12, 119
intercultural research 9, 88, 90, 94–6, 99
International Center for Research on Women (ICRW) 122
International Labor Organization (ILO) 79–82
international migration 57–8, 72, 80
intersecting systems of oppression 67
intersectionality 138, 139, 163, 183

Jamaica 53–67
Janashakthi Bank 38–41, 48–9
Julius Nyerere 143, 146–7

Kabeer, N. 35–6, 45, 47, 123
Kalmunai 186–7, 189–90, 193
Kandyoti, D. 216–17
Kandy District, Sri Lanka 36, 40–1, 43, 45, 50, 187
Katz, C. 3, 56, 88, 93–4, 99
Kawangware-Congo 25–7
Kenya 2, 11, 17–18, 23–4, 28, 142–3, 145, 163–9, 172
Khanga 146–7
Kinniya 186–94
kiosk 24–5
knowledge production 88, 139–41, 145, 164
Kuria, D. 24–6
Kwan, M.P. 123–4, 126, 139, 156, 184
KwaZulu-Natal 103, 106

land 103, 109, 112; ownership 91, 110, 112; reform 5–6, 11, 92, 105, 110–11; rights 105, 110; use 10, 19, 103, 108–10, 113–16, 119, 134, 139, 174
Langata Cemetery 27
liberation struggle 10, 138, 139–44, 148–9, 151, 153, 156
Limpopo Province, South Africa 87, 89–93, 96, 98–9
livelihood 2, 7, 88, 104, 118, 163–5, 169, 171, 173, 182, 188, 194, 201, 211, 215–16; gender 2, 9, 90, 98, 100; economic 4–5, 11, 92–3; subsistence 87
loans 21, 37–41, 43, 45, 183, 211, 214
low-income areas 25, 28

Maltzahn, R. von and Riet, M. van der 106, 108
map 108–9, 112, 114, 129, 134, 139, 149, 154, 167, 181; mental 10, 103, 106, 188; participatory 10, 108, 114, 123, 129, 188, 195, 204; gender 182, 186, 189, 193
Masai 11, 163, 165, 169–72, 174, 176, 177
Maathai, Wangaari 155
microcredit 5, 8, 36–8, 41, 43, 45, 47–50, 125, 217, 219
micro-enterprise 7, 39, 41, 50, 92
microfinance 6, 8, 35–50, 201
migration: migration circuits 54–5, 57; migrant networks 53; West India 62

Mlama P. O. 141, 147
mobility 4–5, 8, 10, 31, 35–7, 43–5, 47–50, 55, 73, 92, 97, 122, 124–6, 129, 132–5, 173, 181, 194–5
Mohanty, C. 2–5, 138, 139–41, 144
Momsen, J. 2, 55, 122, 134, 183–4
Muslim 139, 144, 187–8, 206, 211, 213

Nagar, R. 3, 87, 97
Nagar, R. and Swarr, A. L. 3–4, 89
Nairobi 8, 17–18, 22–31, 135; Central Business District Association 23; City Council 22–4
Nanda Devi National Park 206, 212, 216, 218
natural: disaster 7, 10, 181–6, 196; resources 2, 5–6, 10, 91, 108–10, 201, 203, 216
neoliberal 17, 20–1, 26, 30–1; globalization 1–2, 4–5, 7–9, 18, 91
networks 9, 45–7, 49–50, 59, 156; diasporic 62; knowledge 54; migrant 53; social 93, 175, 219; transnational 56, 148
Nkrumah, K. 142–3
non-borrowers 36, 41–6, 49
non-refoulement 76

Oberhauser, A. M. 92–3
Ogra, M. 201, 203, 217–18

pairwise ranking 108, 111
Pan-Africa 138, 142
participant observation 41, 205
participation: economic 124; labor force 91; labor market 56, 61; microcredit 47, 50; women's 35, 49, 109, 214–15; 217
participatory: gender mapping 11, 186, 189; methods 109, 188–9, 193
Participatory Geographic Information Science (PGIS) (See GIS)
pastoralism 165, 168–70, 173, 212
patriarchy 2, 7, 10, 78, 105, 111, 117; bargains 217; institutions 118
people-park conflict 202
peri-urban 103–6, 110, 112, 118–19
policy makers 8, 19, 22, 53, 60–2, 73, 122, 196
post-apartheid 91–2, 110, 119, 142–3
postcolonial state 147
post-conflict society 155
post-disaster 11, 181–3, 186, 188–9, 192, 195–6

post-independence 22, 146–8, 153
poverty 21, 35–6, 41, 43, 58–9, 62, 92,
     105, 112, 123–6, 152, 169; alleviation
     37–8, 48, 56, 203; reduction 61, 103,
     125
power relations 2–4, 7, 9–10, 18, 42,
     88–90, 99, 104, 151, 171, 182
Pratt, G. 2–3, 57, 87, 97
praxis 1–4, 9, 87–8, 90, 93, 95, 97, 99–100
private sector 17–18, 20–1, 23, 26, 29–30,
     63
privatization 5, 8, 17–21, 23, 25–6, 29–30
problem ranking exercises 10, 103, 106
prostitution 72, 75–80
Protected Area 11, 200, 204, 206–8, 215,
     218–19
public: private partnerships 26, 30–1;
     toilets 8, 18–31

qualitative: data 103, 123–4, 129, 135, 167,
     193; methods/research 10–11, 103–6,
     118–19, 204, 219
quality of life 39, 61, 125

Rajaji National Park 207–8, 213
Rankin, K. 35–6, 45, 49
rehabilitation 11, 181–9, 193–6;
     post-disaster rehabilitation 11, 182,
     186, 188–9, 195
relocation 190–2, 200
remittance 53–4, 56, 59, 61–2, 71–4, 80–1
repayment 41
resource 1–2, 4–7, 10–11, 36–7, 42, 47–9,
     54, 58, 88–9, 91, 93, 97–100, 103–5,
     108–10, 112, 119, 125, 147, 151, 167,
     174, 181–4, 188–90, 194–5, 200–3,
     211–14, 216, 219
Rich, A. 1, 4
Rocheleau, D., Thomas-Slayter, B. and
     Wangari, E. 2, 6, 10, 167, 200–1
Roy, A. 37, 42, 125
rural women 36, 43–4, 89, 92, 94, 98,
     122–6, 128, 132, 134

Samarasinghe, V. 71, 73–4, 80
Samurdhi Bank 38–45, 49
sanitation 8, 17–31, 194; facilities 18–19,
     22; infrastructure 18, 22; services
     17–18, 20–2, 28, 30–1; water and
     17–18, 21–3, 194
Sarvodaya Economic Enterprise
     Development Services (SEEDS)
     38–41, 48–9

self-esteem 42, 44, 47–8, 50
self-help groups 207
sex trafficking 74; victims 71, 72, 74–80
single parenting 57
slum areas 22, 27–9
small business 38–9, 43, 97, 110, 114
social: change 1, 4, 11, 182; construction
     106, 164; construction of gender 2, 9;
     entrepreneurship 24, 26, 30;
     hierarchies 109; reproduction 17, 20,
     56, 58–62; -spatial context 35–7
South Africa 9–11, 19, 21, 87–8, 90–2,
     94–5, 97–100, 103–7, 110–19, 124,
     143, 148–9, 151, 155
Soweto 150
space: boundaries 134; budget 134;
     mobility 8, 31, 45, 50, 124–6;
     relationship 130
Spivak, G. 144
Sri Lanka 8, 11, 35–50, 181–2, 185–96
Stockholm Environmental Institute (SEI)
     167–8
structural adjustment 59
subsistence livelihoods 87
sustainable development 10, 134, 183,
     218–19

Tamil 186–8
Tanzania 10, 138, 139–43, 145–9, 151–3,
     155–6, 165; African National Union
     of Tanzania (TANU) 146, 152
technocratic 184, 189
tenure 22, 104–5, 170, 173, 176
time: geography 124, 126, 135; poverty 7,
     123, 125, 134–7
toilet: malls 22, 24–5, 27–8, 30; studies 17
topographical contour lines 88, 93
trackers 122, 129–30, 132
traditional authority 104, 106
transformative process 7, 11, 219
transnational 1–6, 8–11, 21, 66, 71–3,
     87–100, 139, 148, 156; circuits 56–8,
     65; exchanges 53; family care 57–8;
     feminism xv, xvi, 2–6, 9–10, 87–90,
     93, 97, 99–100; feminist praxis 3;
     migration 8–9, 72–3; networks 56,
     148; spaces 3, 5
triple burden 59
tsunami 11, 181–2, 185–91, 193–6

Uganda 10, 122–135, 155
Ujamaa 143
United Nations Millennium Development
     Goals 8, 18

United Nations Anti-Sex Trafficking
Protocol 8–9, 71–80
urban political ecology 17, 20–2, 30–1
user-fees 25

Venn diagrams 10, 103, 106, 108–9, 114,
117–19
village institutions 202
violence: 46, 58–9, 63–4, 66–7, 77, 89,
144, 148, 151, 155, 188; domestic 46,
63; gender-based 1, 8
Virtual Freedom Trail Project (VFTP) 10,
139, 149, 154, 156

vulnerable 56, 78, 92, 99, 110–11, 165,
183, 186, 192, 200

Walker, C. 92, 103–4
Water and Sanitation Program-Africa
(WSP-Africa) 23
water services 17, 30
West Virginia 91
Western feminism 1, 138, 141, 143
Women's Saving and Credit Group
(WSCG) 207, 211

Young, I. 89–90, 95, 99
youth 64, 126, 194, 218